Daniel J. Vogt, Joel P. Tilley, Robert L. Edmonds
Soil and Plant Analysis for Forest Ecosystem Characterization

Ecosystem Science and Applications

Editors
Jiquan Chen
Heidi Asbjornsen
Kristiina A. Vogt

Daniel J. Vogt, Joel P. Tilley, Robert L. Edmonds

Soil and Plant Analysis for Forest Ecosystem Characterization

—

DE GRUYTER

Higher
Education
Press

This work is in the *Ecosystem Science and Applications* co-published by Higher Education Press and Walter de Gruyter GmbH.

ISBN 978-3-11-055450-2
e-ISBN (PDF) 978-3-11-029047-9
e-ISBN (EPUB) 978-3-11-038176-4
ISSN 2196-6737

Library of Congress Cataloging-in-Publication Data
A CIP catalog record for this book has been applied for at the Library of Congress.

Bibliographic information published by the Deutsche Nationalbibliothek
The Deutsche Nationalbibliothek lists this publication in the Deutsche Nationalbibliografie; detailed bibliographic data are available on the Internet at http://dnb.dnb.de.

© 2017 Higher Education Press and Walter de Gruyter GmbH, Berlin/Boston
This volume is text- and page-identical with the hardback published in 2015.
Cover image: Julian Weber/Hemera/thinkstock
Printing and binding: CPI books GmbH, Leck

♾Printed on acid-free paper
Printed in Germany

www.degruyter.com

Acknowledgments

The authors would like to thank Dr. Kristiina A. Vogt, Patricia A. Roads and Roxanna Lewis for their editorial help on this book. However, any errors are not due to their editorial help but solely due to the authors.

Acronyms

ADP	Adenosine Diphosphate
AFDM	Ash-free Dry Mass
AM fungi	Arbuscular Mycorrhizal Fungi
ATGC	nucleotide bases: Adenine, Thymine, Guanine, Cytosine
ATP	Adenosine Triphosphate
BLAST	Basic Local Alignment Search Tool program for finding regions of local similarity between sequences
CB	Citrate-Bicarbonate
CDB	Citrate-Dithionite-Bicarbonate
cDNA	complementary DNA (cDNA) is DNA synthesized from a messenger RNA (mRNA) template in a reaction catalyzed by the enzymes reverse transcriptase and DNA polymerase, and is often used to clone eukaryotic genes in prokaryotes
CEC	Cation exchange capacity
CFV	Coarse Fraction Volume (as in the > 2 mm particles in the bulk soil)
CTAB	Cetyltrimethylammonium Bromide—a positively charged detergent
CV	Coefficient of Variation-statistical parameter that is the ratio of the standard deviation to the mean
dATP	Deoxyadenosine triphosphate (dATP)—a nucleoside triphosphate used in cells for DNA synthesis (or replication)
DBH	Diameter at breast height (1.3 m) of a tree
dCTP	Deoxycytidine triphosphate (dCTP)—a nucleoside triphosphate that contains the pyrimidine base cytosine
dsDNR	double stranded Deoxyribonucleic Acid (DNA)
DDI	Distilled De-Ionized (water)
DDW	Double Distilled Water
DEA	Dentrification Enzyme Activity
dGTP	Deoxyguanosine triphosphate (dGTP)—a nucleoside triphosphate, and a nucleotide precursor used in cells for DNA synthesis
DNA	Deoxyribonucleic Acid
dNTP	deoxy-nucleotide-tri phosphate—used by the DNA polymerase to add nucleotides to the elongating DNA strand (during replication). dNTP is a generic term referring to the four deoxyribonucleotide triphosphates: dATP, dCTP, dGTP and dTTP

DON	Dissolved Organic Nitrogen
DTPA	Diethylene Triamine Pentaacetic Acid
dTTP	Deoxythymidine triphosphate (dTTP)—one of the four nucleoside triphosphates that are used in the *in vivo* synthesis of DNA
DW	Distilled Water
EC	Electrical Conductivity
ECEC	Effective Cation Exchange Capacity
EDTA	Ethylene Diamine Tetraacetic Acid
EF-1α	Eukaryotic Elongation Factor (also known as EF1A) catalyzes aminoacyl-tRNA binding by the ribosome during translation
EM	Ectomycorrhizas or ectomycorrhizal (fungi or root)
EMF	Electromotive Force (or Voltage)
EPA	The United States Environmental Protection Agency
EtBr	Ethidium Bromide
FAA	Formalin Acetic Acid
FAO	Food and Agriculture Organization of the United Nations
FF	Forest Floor, or sometimes litter or surface organic horizon in forests
FID	Flame Ionization Detector
GC	Gas Chromatograph
GC-rich	Guanine-Cytosine rich means that the GC as a percentage of nitrogenous bases on a DNA molecule is high (from a possibility of four different ones also including adenine and thymine)
GELCOMPAR II	Software Package—a program that makes it possible to link multiple electrophoresis fingerprints to the strains or samples studied and generate multiphasic groupings and identifications with databases of unlimited size
I.D.	Inner Diameter
IC	Ion Chromatography
ICP	Inductively Coupled Plasma Spectroscopy
ID	Identification Number
IRGA	InfraRed Gas Analyzer system
ITS	Internal Transcribed Spacer refers to a piece of non-functional RNA situated between structural ribosomal RNAs (rRNA) on a common precursor transcript
LCI	Ligno-cellulose Index
L-DOPA	L-3, 4-dihydroxyphenylalanine, a chemical that is made and used as part of the normal biology of humans, some animals and plants
LECO CHN	LECO corporation equipment that analyzes Carbon, Hydrogen, Nitrogen (CHN)
LOI	Loss on Ignition
LR0R/LR16	Small conserved subregion for the large subunit (LSU) rDNA gene
LSU rRNA	Large Sub Unit of the ribosomal ribonucleic acid (rRNA)
ML5/ML6	Conserved primer sequence for PCR amplification
MPN	Most Probable Number
NIST	National Institute of Standards and Technology

Oa Horizon	Organic horizon composed of highly decomposed organic material-humus or muck (can see very few signs of the original organic tissues) (USDA Soil Taxonomy [31])
OD	oven dried
Oe layer	Organic horizon composed of intermediate decomposed organic material-peat (can see some signs of the original organic tissues) (USDA Soil Taxonomy [31])
Oi Horizon	Organic horizon composed of slightly decomposed organic material (can still see evidence of the original organic tissues) (USDA Soil Taxonomy [31])
OM	Organic Matter
OSHA	United States Occupational Safety and Health Administration
PCB	Polychlorinated Biphenyl
PCR	Polymerase Chain Reaction
PDAB	p-dimethylaminobenzaldehyde
pH	The decimal logarithm of the reciprocal of the hydrogen ion activity in a solution
PHYLIP	PHYLogeny Inference Package—a free computational phylogenetics package of programs for inferring evloutionary trees
PLFA	PhosphoLipid ester-linked Fatty Acids
PVA	Polyvinyl Alcohol
PVC	Polyvinyl Chloride—rigid forms of PVC are frequently used in the construction of pipes or tubes
QA/QC	Quality Assurance/Quality Control
QC	Quality Control
QF-PCR	Quantitative Fluorescent Polymerase Chain Reaction
qPCR or QPCR	real-time quantitative Polymerase Chain Reaction
QRT-PCR	Quantitative Reverse-Transcriptase Polymerase Chain Reaction
rDNA	Ribosomal Deoxyribonucleic Acid
RFLP	Restriction Fragment Length Polymorphism
RNA	Ribonucleic Acid
RPB2	A gene that encodes for RNA polymerase
RQ-PCR	See RT-PCR
rRNA	ribosomal ribonucleic acid is the RNA component of the ribosome and is essential for protein synthesis in all living organisms
%RSD	Relative Standard Deviation, or also referred to as the absolute value of the Coefficient of Variation (CV)
RT-PCR	Reverse Transcription Polymerase Chain Reaction (not to be confused with Real Time PCR or qPCR)
SI unit	The International System of Units (abbreviated SI from French: Le Système international d'unités) is the modern form of the metric system
SOM	Soil Organic Matter
T	Temperature Correction Factor
TCA	Trichloroacetic Acid

TCD   Thermal Conductivity Detector
TDR   Time-Domain Reflectometry
TDS   Total Dissolved Solids
TE   commonly used buffer solution in molecular biology consisting of
                Tris (pH buffer) and EDTA (cation chelate); used to solubilize DNA
                or RNA while protecting it from degradation
TKN   Total Kjeldahl Nitrogen (Total Nitrogen using the Kjeldahl method)
TNPP   Total Net Primary Production
UPGMA   Unweighted Pair Group Method with Arithmetic mean
USDA   United States Department of Agriculture
UV   Ultra-violet radiation
W-B   Walkley-Black carbon
wt.   Weight

Contents

Part I   Context of Soil and Plant Analysis —— 1

1   Overview of Soil and Plant Analysis for
        Forest Ecosystems —— 3
1.1   Soils are Physically, Chemically and Biologically Complex —— 8

Part II   Introductory Methods in Soil and Plant Analyses —— 15

2   Field Characterization of Soils to Establish
        Sampling Protocols —— 17
2.1   Soil Sampling Design and Methods —— 17
2.1.1   Introduction to Sampling Design —— 17
2.1.1.1   Accuracy, Bias, and Precision —— 18
2.1.1.2   General Considerations on Soil Sampling —— 20
2.1.1.3   Common Sampling Tools and Techniques —— 20
2.1.1.4   Soil Sample Preparation —— 21
2.1.2   Soil Sample Process Procedure —— 23
2.1.2.1   Sample Variability—Number of Samples Required —— 24

3   Plant Tissue Characterization —— 29
3.1   Tissue Sampling —— 30
3.2   Tissue Preparation and Laboratory Extraction —— 33

4   Introduction: Laboratory Practices —— 35
4.1   General Laboratory Protocol —— 35
4.1.1   Safety —— 36
4.1.2   Laboratory Water —— 37
4.1.3   Clean-up —— 37
4.1.4   Waste Disposal —— 39

Part III   Soil Physical, Chemical and Biological Analyses —— 41

5   Methods for Analyzing Soil Physical Characteristics —— 43
5.1   Soil Moisture —— 43

5.1.1   Direct Methods of Estimating Soil Moisture and
            Soil Water Potential —— 46
5.1.2   Procedure to Determine Gravimetric Water Concentrations —— 47
5.2   Soil Bulk Density —— 48
5.2.1   Soil Bulk Density Methods —— 49
5.3   Soil Texture (Particle Size Analysis or Mechanical Analysis) —— 50
5.3.1   Soil Texture Methods —— 53
5.3.1.1   Soil Texture Procedure: Bouyoucos Hydrometer Method —— 56
5.4   Soil Water Potential —— 58
5.4.1   Pressure Plate Apparatus Procedure: Soil Moisture Release Curve —— 63

6   Soil Chemical Characterization —— 67

6.1   Soil pH —— 70
6.1.1   Measuring pH —— 71
6.1.2   The Care of pH Electrodes —— 74
6.2   Electrical Conductivity (EC) —— 75
6.2.1   Saturated Paste Extract Procedure: Electrical Conductivity —— 80
6.3   Ion Exchange in Soils —— 82
6.3.1   Cation Exchange Capacity —— 82
6.3.2   Exchangeable Cations —— 86
6.3.3   Extraction Procedures for Exchangeable Cations and
            Cation Exchange Capacity —— 87
6.4   Exchangeable Soil Acidity —— 91
6.4.1   Extraction Procedures for Exchangeable Soil Acidity —— 92
6.4.1.1   Exchangeable Acidity (Barium Chloride—Triethanolamine
            Method) —— 92
6.4.1.2   Exchangeable Acidity (Potassium Chloride Method) —— 93
6.5   Extractable Inorganic Soil Nitrogen —— 95
6.5.1   Extraction Methods for Inorganic Soil Nitrogen —— 96
6.5.1.1   Single Extraction Procedure: Extractable Inorganic Nitrogen —— 97
6.5.1.2   Double Extraction Procedure: Mechanical Vacuum Extractor —— 98
6.6   Soil Phosphorus —— 100
6.6.1   Methodology for Measuring Soil Phosphorus —— 102
6.6.2   Procedure: Extractable Inorganic Phosphorus —— 108
6.7   Soil Carbon and Organic Matter —— 110
6.7.1   Dry Combustion Procedure: Total Soil Carbon and Nitrogen —— 112
6.7.2   Loss on Ignition (LOI) Procedure: Total Soil Organic Matter —— 113
6.7.3   Walkley-Black Procedure: Soil Carbon —— 114
6.8   Selective Dissolution of Iron and Aluminum —— 116
6.8.1   Extraction Procedure: Organically Complexed Iron and
            Aluminum —— 117
6.8.2   Extraction Procedure: Non-crystalline Soil Iron and
            Aluminum Oxides —— 118

7 Total Plant and Soil Nutrient Analysis (Digestion) ——121

7.1 Wet Oxidation Method —— 121

7.2 Dry Oxidation Method —— 123

7.3 Total Dissolved Carbon and Nitrogen in Water —— 125

7.4 Modified Kjeldahl Digest Procedure: Sulfuric Acid Digest for "Total" Nutrients —— 126

7.5 "Total" Nutrient Analysis Procedure: Dry Ashing Followed by Nitric Acid Digest —— 129

7.6 Total Dissolved Nitrogen in Water Procedure: Persulfate Oxidation —— 130

8 Soil Biology Characterization ——133

8.1 Soil Microbes —— 134

8.1.1 Archaea and Bacteria —— 135

8.1.2 Fungi —— 137

8.1.3 Soil Algae and Cyanobacteria (Blue-green Algae) —— 138

8.2 Methods for Determining Soil Microbial Diversity and Populations—Numbers and Biomass —— 139

8.2.1 Direct Culture, Microscopy and Image Analysis —— 139

8.2.2 Microbial Numbers and Microbial Biomass —— 139

8.3 Mycorrhizas —— 140

8.3.1 Types of Mycorrhizas —— 140

8.3.2 Sampling Mycorrhizas —— 142

8.3.2.1 Sampling Design —— 143

8.3.2.2 Collection of Root and Soil Samples —— 143

8.3.2.3 Storage of Samples —— 143

8.3.2.4 Determining Mycorrhizas in Samples —— 144

8.3.3 Determination of Mycorrhizal Fungal Species —— 144

8.3.3.1 Analysis of Sporocarps and Spores —— 145

8.3.3.2 Morphotypes of Ectomycorrhizas —— 145

8.3.3.3 Trap Cultures for Arbuscular Mycorrhizal Fungi —— 145

8.3.3.4 DNA and Biochemical Techniques —— 146

8.3.4 Ectomycorrhizal Quantification —— 147

8.3.5 Identification of Ectomycorrhizal Sporocarps —— 150

8.3.6 Quantification of Arbuscular Mycorrhizal Colonization —— 151

8.3.6.1 Staining —— 151

8.3.6.2 Grid-line Intersection Method with a Dissecting Microscope —— 152

8.4 Indirect Indices for Soil Biological Activity —— 154

8.4.1 Soil Respiration —— 154

8.4.1.1 CO_2 Gas Sampling —— 155

8.4.1.2 The Soda Lime Technique —— 157

8.4.2 Decomposition Rates of Litter —— 159

8.4.2.1 Fine Litter —— 159

8.4.2.2 Woody Debris —— 161

8.4.2.3 Fine Woody Debris —— 161

8.4.2.4   Coarse Woody Debris (CWD) —— 163
8.4.2.5   Standard Substrates —— 165
8.4.2.6   Calculation of Decomposition Rates —— 167
8.4.3   Soil Enzymes —— 168
8.4.4   Functional Biodiversity—Phospholipid Ester-linked Fatty Acids (PLFA)
            and Substrate Utilization Profiles —— 169
8.4.5   Molecular Tools for Ecological Systems —— 170
8.4.5.1   DNA Extraction from Soil —— 171
8.4.5.2   PCR —— 172
8.4.5.3   Restriction Length Fragment Polymorphism RFLP —— 174
8.4.5.4   Primers —— 174
8.4.5.5   Gel Electrophoresis —— 174
8.4.5.6   DNA Sequencing —— 175
8.4.5.7   Metagenomics and Transcriptomics —— 176
8.5   Soil Invertebrates —— 176
8.5.1   Macrofauna with Emphasis on Earthworms —— 178
8.5.1.1   Extraction Methods —— 178
8.5.2   Mesofauna —— 180
8.5.2.1   Mites and Collembola —— 180
8.5.2.2   Enchytraeids —— 183
8.5.3   Microfauna —— 184
8.5.3.1   Protozoans —— 184
8.5.3.2   Rotifers —— 184
8.5.3.3   Tardigrades —— 184
8.5.3.4   Microfauna-Nematodes —— 185
8.6   Nitrogen Transformations —— 186
8.6.1   Nitrogen Fixation —— 187
8.6.1.1   Acetylene Reduction —— 188
8.6.1.2   N Accretion through Time —— 190
8.6.1.3 ^{15}N Based Methods —— 190
8.6.2   Denitrification —— 191
8.6.2.1   The Acetylene Inhibition Method —— 192
8.6.2.2 ^{15}N Tracer Methods —— 192
8.6.2.3   Direct N_2 Quantification —— 192
8.6.2.4   Mass Balance Approaches —— 192
8.6.2.5   Stable Isotope Approaches —— 193
8.6.2.6   Approaches using *in situ* Gradients in Environmental Tracers —— 193
8.6.2.7   Molecular Approaches —— 193

Appendices —— 195

References —— 203

Subject Index —— 217

List of Tables

Tab. 1.1   Processes that form all soils —— 10
Tab. 1.2   Processes that form all soils but depend on environmental conditions —— 10
Tab. 1.3   Factors that cause soils to degrade —— 11
Tab. 2.1   Definitions of some sample measurement terms —— 19
Tab. 2.2   An example of calculating the variability and precision associated with sampling —— 25
Tab. 5.1   United States Department of Agriculture (USDA) textural classes of soils based on the USDA particle-size classification and potential ranges of percentages of soil separates —— 53
Tab. 5.2   Water potential in the soil-plant-atmosphere system —— 61
Tab. 6.1   Role of some mineral elements in plant growth and functions and mineral links to human health —— 68
Tab. 6.2   Some nutrient concentration ranges in plants and soils —— 69
Tab. 6.3   Descriptive terms for ranges of soil pH —— 74
Tab. 6.4   Definition of Soil Salinity classes related to Electrical Conductivity (EC) —— 76
Tab. 6.5   Soil types related to pH, EC, SAR, and ESP classes —— 76
Tab. 6.6   Effects of Electrical Conductivity (EC) on crops —— 77
Tab. 7.1   Percent recovery of 11 elements in a modified Kjeldahl digest ——122
Tab. 7.2   Suggested calibration series for Dissolved Organic Nitrogen (DON) determination ——131
Tab. 8.1   Numbers of species and genera in the bacteria phyla ——136
Tab. 8.2   Characteristics of the different types of mycorrhizas ——141
Tab. 8.3   Example calculation for CO_2 evolution from a 15-cm diameter cylinder in the soil ——159
Tab. 8.4   Calculations for estimating k (yr^{-1}) for decaying Pacific silver fir needles ——168
Tab. C.1   Example of a Field Data Recording Sheet ——198
Tab. C.2   Soil Data Summary Sheet ——200
Tab. C.3   Soil Water Concentration Data Sheet ——200
Tab. C.4   Soil Bulk Density Data Sheet ——201

List of Figures

Fig. 1.1   Factors of human land uses and their influences on our natural ecological systems including soils —— 4

Fig. 1.2   Total net primary productivity in tropical moist forests by soil texture classes —— 13

Fig. 5.1   Texture triangle of different proportions of sand, silt, and clay forming 12 texture classes —— 51

Fig. 5.2   Comparison of particle size scales according to different organizations —— 52

Fig. 5.3   Soil texture classes determined by hand —— 54

Fig. 5.4   Schematic diagram of water molecule —— 58

Fig. 5.5   An example of hysteresis in soil water retention curves of wetting and drying curves —— 63

Fig. 6.1   A schematic diagram of a pH glass electrode and expanded view of the electrode bulb —— 73

Fig. 6.2   A schematic of an Electrical Conductivity Cell —— 78

Fig. 8.1   Using the gridline intersection method for estimating total root length and numbers of ectomycorrhizal root tips ——149

Fig. 8.2   A gridline intersection example using an 8.5 cm diameter round Petri dish with a 1/2 inch (1.27 cm) grid and a 1 m test sample of thread cut into fragments and randomly re-distributed 10 times ——153

Fig. 8.3   Diagram showing examples of calculating different volumes of a log such as total, wood and bark volumes ——164

Fig. 8.4   Average ash-free dry mass of Pacific silver fir needles remaining in litter bags at five sampling times in the Cascade Mountains of Washington, USA ——168

Fig. 8.5   A schematic of the Tullgren or Berlese funnel to extract soil invertebrates from the soil ——181

Fig. 8.6   High gradient invertebrate extractor ——182

Fig. A.1   Frequency distribution of 100 normally distributed random observations, with mean = 99.582 and standard deviation = 9.773. ——195

Fig. A.2   Frequency distribution of 20 means of 5 observations each ——196

Fig. A.3   Frequency distribution of 10 means of 10 observations each ——197

List of Equations

Eq. 1.1   Soil and it's properties —— 9

Eq. 2.1   Percent rock —— 24

Eq. 2.2   Variance (s^2) and standard deviation (s) —— 24

Eq. 2.3   Coefficient of variation (CV) —— 25

Eq. 2.4   Sample number (n) for statistical sampling —— 26

Eq. 5.1   Soil water concentration (w_d)—gravimetric dry-weight basis —— 44

Eq. 5.2   Soil water concentration (w_m)—gravimetric wet-weight basis —— 44

Eq. 5.3   Soil water concentration (w_d) calculated using w_m —— 44

Eq. 5.4   Soil water concentration (w_m) calculated using w_d —— 44

Eq. 5.5   Soil water concentration (w_v)—volumetric basis —— 44

Eq. 5.6   Soil water concentration (w_v) calculated using w_d, soil bulk density and water density —— 45

Eq. 5.7   Soil water concentration (w_f) calculated using w_v and depth —— 45

Eq. 5.8   Soil water concentration (w_d) calculated using moist soil and dry soil weights —— 45

Eq. 5.9   Soil water concentration (w_m) calculated using moist soil and dry soil weights —— 45

Eq. 5.10   Dry soil weight calculated using moist soil weight and w_d —— 45

Eq. 5.11   Dry soil weight calculated using moist soil weight and w_m —— 45

Eq. 5.12   Moist soil weight calculated using dry soil weight and w_d —— 45

Eq. 5.13   Moist soil weight calculated using dry soil weight and w_m —— 45

Eq. 5.14   Soil bulk density (ρ_b) calculated using volume fraction of air pores, soil mineral particles, and organic matter —— 48

Eq. 5.15   Percent pore space calculated using soil bulk density and mineral particle density —— 48

Eq. 5.16   Coarse Fraction Volume (CFV) calculated using coarse fraction mass and particle density —— 49

Eq. 5.17   Soil particle velocity (v), i.e., sedimentation, calculated using Stokes' Law —— 54

Eq. 5.18   Percent sand calculated using the Bouyoucos hydrometer method —— 57

Eq. 5.19   Percent clay calculated using the Bouyoucos hydrometer method —— 57

Eq. 5.20 Percent silt calculated knowing the percent sand and clay —— 57

Eq. 5.21 Pressure difference (ΔP) across the air/water interface knowing the surface tension and curvature radius of the water —— 62

Eq. 6.1 Water and its dissociation —— 70

Eq. 6.2 Ion product for water (K_w) —— 70

Eq. 6.3 Definition of pH —— 70

Eq. 6.4 Al^{3+} hydrolyzing in water releasing a hydrogen ion increasing soil acidity —— 71

Eq. 6.5 Relationship between Electromotive Force (EMF) and pH at 25°C governed by the Nernst equation —— 73

Eq. 6.6 Calculation of an estimate of total cation (or anion) concentration —— 80

Eq. 6.7 Calculation of an estimate of Total Dissolved Solids (TDS) —— 80

Eq. 6.8 Calculation of an estimate of osmotic potential (Ψ_π) —— 80

Eq. 6.9 Calculation of base saturation —— 86

Eq. 6.10 An example of converting an analytical concentration to equivalent units —— 88

Eq. 6.11 Calculation of the mass and volume of the entrained solution —— 90

Eq. 6.12 Conversion of the concentration in the extract to total mass —— 90

Eq. 6.13 Calculation of Cation Exchange Capacity (CEC) —— 90

Eq. 6.14 Calculation of Organic Phosphorous using the extraction method —— 104

Eq. 6.15 Calculation of Organic Phosphorous using the ignition method —— 104

Eq. 6.16 Reaction of adding ^{32}P as orthophosphate ions to a soil-water system —— 107

Eq. 6.17 Calculation of ^{31}P solid from knowing ^{32}P solid, ^{32}P solution, and ^{31}P solution —— 107

Eq. 6.18 Chemical reaction for determining the Walkley-Black readily oxidizable carbon —— 112

Eq. 6.19 Calculation of $FeSO_4$ concentration in the Walkley-Black carbon analysis —— 115

Eq. 6.20 Calculate the mass of reduced chromium in the Walkley-Black carbon analysis —— 115

Eq. 6.21 Calculate the Walkley-Black % Carbon —— 115

Eq. 6.22 Calculate a Walkley-Black recovery factor and Total Soil Carbon —— 116

Eq. 8.1 Calculation of percent Arbuscular Mycorrhizal (AM) root length —— 153

Eq. 8.2 Ideal Gas Law to calculate concentration mass, e.g., for CO_2 —— 156

Eq. 8.3   Calculation of gas flux, e.g., CO_2 — 156
Eq. 8.4   Calculation of soil CO_2 evolution using the Soda Lime
           method — 158
Eq. 8.5   Calculation of the volume of a cylinder — 164
Eq. 8.6   Calculation of the volume of a fragmented layer of a cylinder — 165
Eq. 8.7   Calculation of the volume of the elliptical decomposed log — 165
Eq. 8.8   Calculation of negative exponential model of litter mass
           remaining over time — 167
Eq. 8.9   Calculation of the contaminant-corrected ethylene concentrations
           in the acetylene reduction method — 189
Eq. 8.10   Calculation of the ethylene production rate in the acetylene
           reduction method — 189
Eq. 8.11   Calculation of the acetylene reduction activity in the acetylene
           reduction method — 189

Part I
Context of Soil and Plant Analysis

Chapter 1

Overview of Soil and Plant Analysis for Forest Ecosystems

Humans have become one of the greatest engineers of change to the land and the environment. No other organism has altered its degrees of freedom as drastically as humans. In fact, humans have altered up to 80% of the temperate broadleaved forests and from 25% to 46% of the tropical forests [1]. Today, we are asking many questions regarding environmental issues. These issues include: where and how we can detect, monitor, or measure the impacts of land-use changes or the conversion of land archaeas to different vegetation communities and what will be their effects on our environment. These human land-uses changes have triggered many alterations to soil conditions or its health and have also affected what vegetative community will be able to grow or dominate in a particular landscape (see Fig. 1.1).

The connection between forests and soils is especially apparent from looking at the factors that caused past human societies to collapse. Most of the examples that exist are a result of each society altering the soils chemical, biological or physical characteristics due to resource over-exploitation, e.g., intensive deforestation or intensive and long-term agricultural practices. These land-use activities were followed by a significant change in climate, usually a drought or excessive rainfall, which caused the collapse of the society. It is important to note that generally these negative changes in the soil were only recognized to have contributed to societal collapse after some unusual climatic event, such as a mega-drought or excess rainfall. Several recorded examples of societal collapse linked to human's treatment of the land are:

-   More than 4,000 years before the current epoch the Sumerians of Mesopotamia (modern day Iraq and portions of Syria, Turkey and Iran) cut down cedar forests in the mountains to build their civilization. The hillsides and mountainous areas were bared and the salt-rich sedimentary rocks of the north rapidly eroded into their agricultural fields. This erosion, plus the salts that were transported to their lands, decreased agricultural production by 42% [3,4]. This is the first recorded case of erosion resulting from land-use activities contributing to the collapse of a civilization [5].

Dynamic Land Transitions

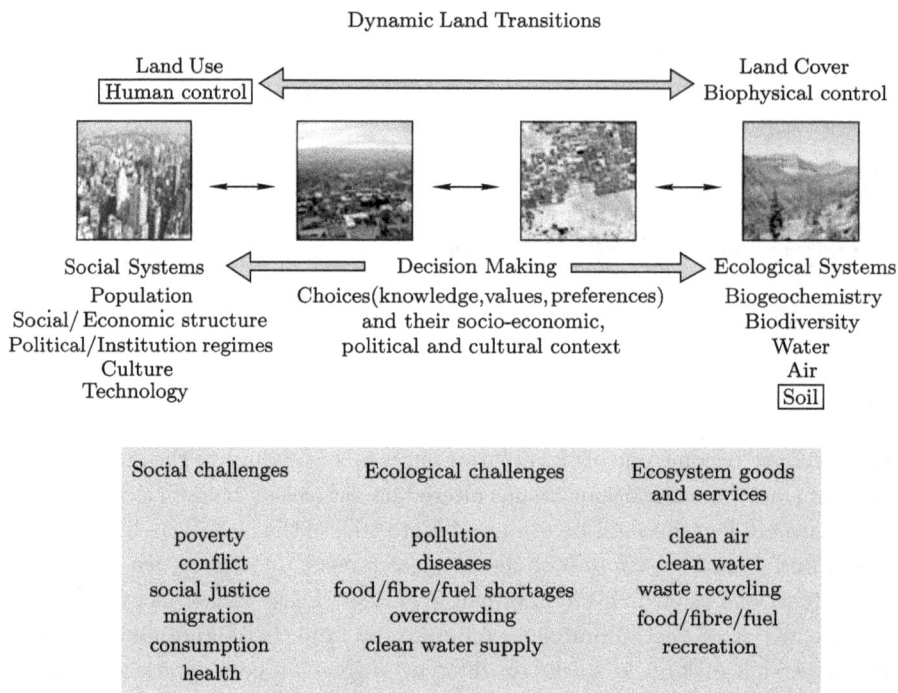

Fig. 1.1 Factors of human land uses and their influences on our natural ecological systems including soils (modified from Global Land Project [2]).

– The Nazca people of Peru lived in a dry forest area dominated by *Prosopis pallid*. This tree played a vital role in maintaining the habitat because its extensive roots pumped water from deep aquifers to the surface of the ground. To expand their agricultural fields, they cut down the trees. This deforestation was followed by a mega El Niño event that hit the southern coast of Peru in about 500 CE and brought excess precipitation. The Nazca's agricultural fields were washed away because the soil's stability was lost without the stabilizing effect of the *P. pallid* forests [6].

– The civilization living on Easter Island (off the western coast of South America) collapsed around 900 CE. One theory suggested that they also collapsed because of deforestation that caused soil erosion during rain and windstorms [7]. This deforestation also caused the loss of soil carbon and the leaching of soil nutrients so that agricultural production collapsed [7]. Deforestation was a result of a rivalry between clans based on which clan could sculpt and transport the largest statue to their territory (immense 8 meter high stone statues called moai). According to Diamond [7], trees from their forests were cut to build massive sleds to hold the statues while fibrous tree bark was used to make ropes to pull each statue from the quarry to its intended site. The size of these statues required from 50 to 500 people to move them.

–   The Maya civilization (situated in the southern part of Mexico and large parts of Central America) collapsed between 1,020 and 1,100 CE after an extended period of drought [8]. This was preceded by "anomalously high rainfall, which favored unprecedented population expansion and the proliferation of political centers between 440 and 660 CE" [8]. This period of population expansion and economic development resulted in over-exploitation of their forests to build their economies. By losing their forests, they lost the buffering that forests would have provided them against the drought.

Even if soil degradation wasn't the main causal factor of the societal collapse, degraded soils were frequently a major component of societies not being sustainable. Because of the importance of soils to humans, methodology books have been developed to not only present methods for analyses but also to ensure standardization so that relative and valid assessments can be made. These books describe soil and plant analyses and provide a context wherein all possible assessments are considered to help determine which sampling and analytical methodologies should be used. This context helps explain why soils, plants, their interconnectivities, and their analyses are an important element of our assessment toolkit. Soil and plant connections are not just relevant in human managed landscapes but are also part of connections and processes that occur in the natural environments. The natural environment generally maintains a threshold for nutrient applications to any ecosystem so that luxury levels do not pollute our environments. Soil and plant analyses are therefore useful to warn of ecosystems' approaching the tipping point at which nutrient levels become deficient and will decrease plant growth rates. Thus they can be important explanatory factors or indicate causal drivers of changes that occur in our environment due to multiple human and non-human factors. These analyses can allow humans to assess whether land-use practices may become negative and perhaps even decrease both social and environmental resiliencies. These tools become extremely important given the fact that the Earth has 7 billion plus mouths to feed and its soil health has changed in our attempt to provide the needed food.

One only has to read reports such as the Millennium Ecosystem Assessment [9] put out by the United Nations or the Global Land Project [2] to comprehend the large-scale changes to the environment that are attributed to human activities. These changes are occurring globally. Climate change also aggravates and further reduces soil health and its productivity following the impacts that have already resulted from previous land-use changes. This means that there is a need to acquire a better grasp regarding the links between soil conditions and climate change and disturbances in general. Research at the plot level has shown the links that exist between altered soil conditions and the productive capacity of vegetative communities. However, when comparisons are made at a country level, which is the most common scale used, it is difficult to link land-use changes to the loss of a lands productive capacity. When Vogt *et al.* [10] compared nine indices characterizing a country's environmental and social ranking for 34 countries, few links were identified between the land degradation severity ranking [11] and a countries performance index, its total ecological footprint, its environmental vulnerability index, or its global climate risk index. This seems to indicate that the scale at which soils are included in global assessment may have little value in detecting the loss of environmental or social resilience due to land-use activities. However, this is not correct and merely shows that soils are primarily included in environmental assessments at a scale too coarse to perceive any relationships. To observe the relationships between land-use or climate change and soils, the soil analysis should be conducted at the scale where these interconnectivities can be detected, e.g., soil texture groups. The scale of the disturbance may be less relevant since each soil type will respond differently to a disturbance as well as the vegetative community that grows on that soil. It follows then that soil characteristics can determine the inherent limitations of any land and how that land will respond to climate change disturbances.

It is well accepted that climate change and past land-use activities have already impacted and altered soil health and resilience. The soils are the memory or link to past human land-uses that they have already experienced and this legacy will determine the future ability of humans to acquire resources from that soil. Even though soils may appear to be well buffered from past land-use activities, the soils carry the legacy of these impacts. Today, the intensity of land-use activities continues to increase so that it is even more important to be able to detect when a soil will lose its resilience since this directly impacts the type, amount, and quality of ecosystem services derived from forests and other lands that are utilized by humans.

Many different human activities created the legacies retained in the soil, which contributes to the loss of soil resilience. Changes in soil chemistry have been especially important in leaving a soil legacy that decreases the nutrient supply capacity of a soil. When the western world began to industrialize and combust fossil fuels, the elements released by factories produced acid rain, e.g., rain containing excess nitrogen and sulfur compounds. This acidic rain caused an increase in soil nitrogen levels that leached basic nutrient elements from the soil [12]. These impacts of acid rain became apparent when large scale mortality of trees were noted and published in the media. Stressed and unhealthy trees were a result of trees being unable to acquire basic cations that they needed to grow. The tipping point for these stressed trees followed an extreme drought. These altered nutrient levels became the legacy of acid rain and are difficult to mitigate since it is a complex process to reapply these basic cations into the soil. Despite the fact that trees are not stressed during the non-drought years, these same trees will become stressed when a major drought re-occurs in the regions that previously experienced these tree mortality events. Acid rain was especially problematic since it introduced higher nitrogen levels into soils which native plants were not adapted to utilize. Furthermore, soils with an acid rain legacy and altered chemistry conditions tend to provide conditions suitable for non-native plant species. This means that vegetative communities' changes are frequently a result of the alterations that have occurred in the soils.

These connections between soils and plants are not confined to a particular scale of analysis at which a human land-use occurs. For example, GLP [2] reported how the "Environmental and social dynamics operate across multiple scales, with many connections between the dynamics at different scales. ... Relatively simple approaches to understanding local phenomena in the context of their immediate regions are no longer sufficient." What the figures show are the interconnections between the existing land cover which is biophysically controlled and the ecological system that includes soils, air, water, biodiversity, and biogeochemistry [2] (see Fig. 1.1). In the past the biophysical environment had control over the natural dynamics or cycles of change in nature. Today, the land cover is now being strongly controlled by human land-uses that are creating social and ecological challenges as well as impacting the delivery of ecosystem services. History records many civilizations which have collapsed because of the alterations or the changes that the humans in that society made to the land cover.

To mitigate the impacts of climate change, some of the few options available to society are to increase carbon sequestration in forests and in soils [10]. This is a problem facing industrialized countries where soil carbon levels have decreased by more than half due to long-term agriculture. Most of the developing countries are situated in areas where the climatic conditions, e.g., high rainfall, have already resulted in the leaching of nutrients and loss of soil organic matter. Despite this, most of the developing countries that have forests are able to sequester all of their carbon emissions. The growth rate of these forests is sufficiently high that forests sequester more carbon than is emitted by industries during a year [10]. The high land area in forests explains the large amount of carbon sequestered in forests since the carbon and nutrient poor soils reduces a forests productive capacity. These are also the locations where indigenous communities actively managed their soils to increase their carbon storage capacity, i.e., terra preta soils, more than $1,000$ years ago.

In contrast, the industrialized countries are able to only sequester 20%–50% of their carbon dioxide emitted during fossil energy combustion each year [10]. In the industrialized countries they do need to consider how much carbon is sequestered in their soils since the soils store three-fourths of the terrestrial carbon compared to vegetation [13]. However, soil organic matter accumulations are dependent upon decomposing plant and animal matter. Unfortunately, some of the products of the decomposing material becomes immobilized in biologically unavailable forms on the surface of soil mineral surfaces or aggregate and precipitate out due to inorganic cations [14]. Historical agricultural practices have also decreased soil carbon levels by 33%–50% of what is found in areas where grasses or trees have been growing for a long time [5]. This means that soils are a pool that can be managed to restore carbon storage as organic matter to mitigate climate change.

Deciphering the role of human land-use activities as compared to natural disturbances on ecosystem health and productivity has been difficult. Recently, attention is shifting back to detecting and measuring changes occurring in the soil. Researchers have recognized the importance of soil chemistry or physical properties as tools to determine how resilient are the vegetative communities that grow on each soil type [10]. Even the Natural Resources Conservation Service (NRCS) in the US recently announced that they have developed customized tools that can be used by farmers and ranchers to manage their soils to mitigate climate change impacts [15]. This tool allows each land-use manager to determine if their management practices are, or and not, contributing to removing carbon from the atmosphere. This tool is based on soil characteristics, tillage and the amount of nutrient use, and is based on sampling over 6,000 locations.

Soil characteristics are important in determining how resilient forests are to climate change. This recognition by the global scientific community has resulted in increasing the focus towards comprehending how different soil characteristics connect to plants. It has also focused attention on the impact of human landuses on the changing productive capacity of natural ecosystems such as forests. Soils are one of the critical tipping points that will determine whether vegetation communities can adapt to climate change. Natural or human induced soil constraints will also determine whether ecosystems will continue to provide the ecosystem services that humans are dependent upon for their survival [9].

1.1 Soils are Physically, Chemically and Biologically Complex

Soil components are well described in numerous books and other publication [5, 11, 16] which we encourage you to read. A short summary will be provided since it supports the need to understand how to characterize soils and their methods of analysis. Soils are complex and the appropriate soil methods need to be used or inaccurate results are possible. This explains the reason why methods are included in this book.

Not only the capacity of soils to sequester carbon for longer time scales but also its chemistry and physical properties will determine what the productive capacity or growth rates of forests or even an agricultural crop will be reached [10]. Soil chemistry also determines what kinds of management activities can be used to improve plant growth rates or whether a soil has toxic elements that might decrease plant growth rates. The soil constraints will ultimately determine whether society is able to continue to collect ecosystem services and where agricultural crops can be grown. The fact that about 10% of the total global terrestrial area on average is used for agriculture alludes to the fact that crops cannot be grown everywhere and in many cases because of soil constraints [13, 17]. Land or soil constraints may consist of many factors that have been summarized as: high aluminum levels, salinity, nutrient deficiencies, drought risk due to texture, and erosion risks [11].

Early soil science discussions of soil development have been attributed, in part, to Vasily Dokuchaev in Russia, K. D. Glinka in Germany, and Curtis Marbut, Eugene Hilgard and Hans Jenny in the United States. Their works have contributed to building of a generalized equation that describes why soils are so physically, chemically and biologically complex. This soil development Equation (1.1) also helps to explain the distribution of different soil types around the world and the different chemical, physical and biological properties of soils. The equation is commonly shown as:

Soil and it's Properties

$= $ function of (Climate, Organisms, Relief, Parent Material, Time, ...) (1.1)

The missing factors at the end of this equation generally represent factors that are more localized such as salt inputs from the atmosphere, etc. This equation also shows why there is a high variability in soil attributes and why these attributes are important to measure when determining whether a soil is healthy or resilient to disturbances. They also define what factors will constrain a lands productive capacity [10]. The reason why this knowledge is important is summarized in Vogt *et al.* [10],

> "Soils are very diverse and can be excessively moist or well-drained, shallow or deep to the bedrock, have low to high fertility, and/or have low to high water-holding capacity. All these characteristics will affect its erosion potential; which plants can grow well in it; which plants will successfully out-compete others growing in the same environment; and whether grazing animals (including humans) will derive sufficient nutrients from eating plant materials."

Many of the processes that form soils are also the same processes that human land-use activities alter and have the potential to decrease the productive capacity of a soil or its resilience to other disturbances. The processes that occur to specifically form soils were summarized by Reference [5] and presented in Tables 1.1 and 1.2.

Tab. 1.1 Processes that form all soils [5].

Addition and partial decomposition of organic matter—soil organic matter is naturally derived from the decomposition of plant materials by microbes and soil animals
Formation of structural units—organic matter and clay particles form aggregates in soil that produce good environments for plants to grow

Tab. 1.2 Processes that form all soils but depend on environmental conditions [5].

Leaching and acidification—excess rain and/or natural and anthropogenic acidic rain drains soluble substances from the soil profile
Clay eluviations—clay is washed from the upper profile and is deposited in the lower parts of soil profiles
Podzolization—organic acids or polyphenols from decomposing plant materials complexes with iron, aluminum and clay and is deposited by rain lower in the soil profile
Desilication—greater leaching of silica compared to iron and aluminum resulting in old, nutrient poor soils
Reduction—poor drainage results in soil oxygen being replaced by water
Salinization—accumulation of sulfates and chlorides in soils
Alkalization—accumulation of sodium in soils; these cause soil pH to rise over 8.0, a loss of soil structure that make cultivation of crops difficult
Erosion and deposition of eroded soil—wind and water cause erosion of soils

It also describes why soil qualities vary considerably throughout the world and why some locations have more soil constraints that control plant production or crop yields. It also explains why humans are only as healthy as the delivery capacity of soils that plants grow in and why agricultural managers have to fertilize and irrigate their crops. Each soil will influence the health of humans and other organisms dependent upon eating plants to obtain their nutrients that originally are derived from soils [10].

Food and Agriculture Organization (FAO) identified eight factors that are especially constraining to growing agricultural crops [11]. These eight factors will be mentioned next but it is worth noting that forests are adapted to all of these constraints and particular species of trees are found growing under each of these constraints. Despite trees being adapted to grow under these soil conditions, converting forests into agricultural fields introduces plants that are not adapted to these conditions. In fact, it is not uncommon to find only 30% to 41% of the total land area being arable and good for agriculture. Therefore, productive soils that can be used in intensive crop growth are the exception and not the norm.

The need to recognize and select crops capable of growing on specific soils is a normal part of agriculture and has been practiced by human societies for thousands of years [4]. According to the FAO data summarized by Vogt *et al.* [11] the following eight soil constraints will limit crop growth:

1. Hydromorphy—poor soil drainage
2. Low cation exchange capacity (CEC)—low capacity to retain added nutrients
3. Aluminum (Al) toxicity—elemental toxicity and strong acidity
4. High phosphorus (P) fixation—high levels of ferric and aluminum oxides (in acidic environments) or calcium oxides and hydroxides (in alkaline environments) complex with phosphorus which make it unavailable to plants
5. Vertic properties—clays that contract and expand with moisture changes
6. Salinity and sodicity—presence of salts and sodium (Na) in solution
7. Shallowness—shallow soil depth and the growing medium with rock near the surface (less rooting volume available)
8. Erosion hazard—high risk of soil erosion due to steep slopes or moderate slopes, but also erosion-prone soils due to texture, mineralogy, organic matter, climate, hydrology, etc.

If you compare the list of processes that form soils to the causes of soil degradation, there are several common factors included in both lists. Similar to the FAO study [11], Wild [5] also provided a list of factors that cause soils to degrade and to lose their productive capacity and health. He listed (Tab. 1.3).

Tab. 1.3 Factors that cause soils to degrade [5].

Erosion	Depletion of plant nutrients
Acidification	Reduction of soil organic matter contents
Salinization and sodification	Compaction and crusting
Accumulation of toxic elements	Waterlogging, except for rice

Comparing Wild's [5] summary of the processes that form soils to the soil forming factors shows how humans alter soil forming processes that result in a decrease in soil health and reduced social and environmental resilience. For example, acid rain caused the leaching and acidification of soils that caused the mortality of trees during droughts that occurred in the mid-1970s. The Sumerian civilization collapsed as they lost the productive capacity of their agricultural fields when erosion from deforested mountain areas deposited salts into their agricultural fields. Similarly the Nazca experienced erosion and loss of the lands productive capacity when they converted forests to agricultural fields. Humans have decreased the amount of organic matter that would be naturally added back to soils and the amount of carbon that is stored in soils due to their agricultural practices. This reduction in soil organic matter means that societies have to irrigate and fertilize their agricultural fields since the soils are unable to deliver natural levels of water and nutrients to growing plants.

Examples of societies that collapsed and negatively altered soil quality due to their deforestation is not just something that occurred in the past—it still occurs today. Ehui and Hertel [18] documented the impacts on soil quality due to deforestation in the Cote d'Ivoire. This study showed how the physical and chemical properties of the soil changed due to deforestation, e.g., a 26.9% decline in crop yields resulted from every 10% increase in the cumulative area of forests cut down. These effects were the result of the loss of soil organic matter when the trees were cut. The loss of soil organic matter increased soil erosion. Soil erosion also increased after farmers attempted to grow crops on lands that are marginal for agriculture but are not marginal for forest growth.

Today, deforestation is not the only problems that impacts soil characteristics. In the tropical region, frontier forests are producing a chemical legacy in soils that will have even a more significant impact soil health and productivity, and ultimately human health. Today, the world is experiencing another gold rush that is playing out in forest areas in the American and African tropics that is being driven by the high price of gold on global markets [19]. This activity is already leaving a chemical footprint in the soils. This chemical footprint is being produced by the use of mercury to extract gold from rocks and mercury is toxic to soil health and resilience.

These examples show how soil processes are closely linked to human land-use activities and the need to assess soil processes at the same time as assessing plant biomass and production. The texture diagram with an overlay of total net primary production (TNPP, Mg^{-1} ha^{-1} yr^{-1}) for tropical moist forests (Fig. 1.2) shows one way how production could be constrained by soil physical, chemical and biological properties. This textural triangle shows that generally higher TNPP occurs with finer textures of clays and lower TNPP occurs on sandier textures. Figure 1.2 shows how the soil ability to store water and nutrients needed by plants to grow ultimately determines what growth rates are found for a given piece of land. When TNPP is naturally low, these areas will need supplements of nutrients and water to grow plants at higher growth rates. When these nutrients are applied to agricultural fields, it results in environmental externalities such as pollution [20]. The interest to increase organic farming has been fostered by the idea of reducing the negative environmental impacts of intensive agriculture.

One of the major roles of organic farming is to increase the organic matter of soils so that the land itself can deliver the nutrients and water needed by growing plants rather than artificially saturating the soils with nutrients, much of which frequently leaches from the soils and pollutes the groundwater and/or streams.

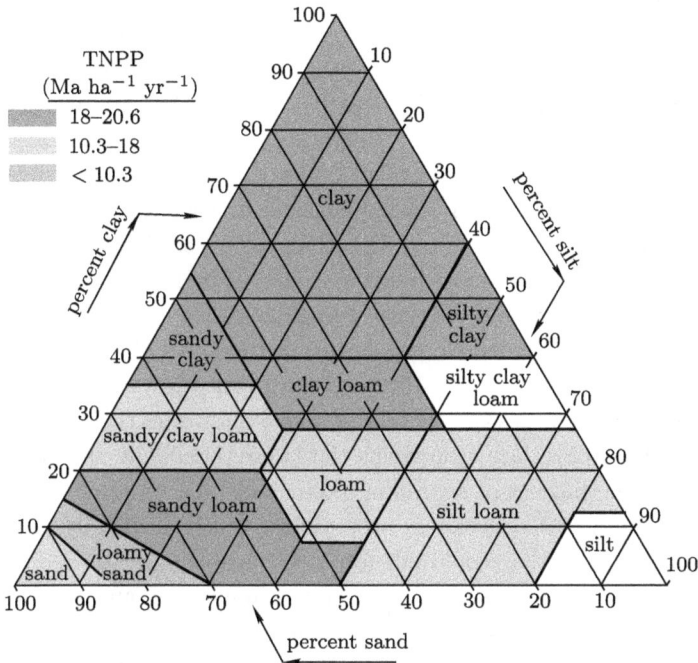

Fig. 1.2 Total net primary productivity in tropical moist forests by soil texture classes [21].

The crops or plants growing on any soil are healthy only as the soil is in providing the nutrients that require to grow. This directly impacts humans and other organisms dependent upon the land for food. The fact that some soils are not as healthy can be seen in how geophagy (a practice of eating soil or clay) is practiced in many parts of the world. For example, in Kenya elephants have been monitored going into caves to eat soil to obtain the sodium they require because the plant tissues they eat contain insufficient amounts for their needs. Animals and even humans acquire other nutrients from eating clay such as phosphorus, potassium, magnesium, copper, zinc, manganese and iron [22].

The indigenous people who live in the Brazilian Amazon recognized the poor quality of their soils more than 1,000 and 2,000 years ago [23]. Their response to this was to increase the organic matter content of soils by adding charcoal to soils. This practice increased soil carbon levels to twice of what is found in an untreated soil. These soils still exist today and are very productive for plant growth. These soils have high nitrogen, phosphorus, calcium, potassium, magnesium, manganese, zinc, and copper levels which are all needed by growing plants [24]. These soils also have high microbial activity and are less prone to the leaching of nutrients which is important in these high rainfall areas [23].

It is clear that there are strong interconnectivities with human land-use activities (e.g., deforestation, urban development, some intensive agricultural and forestry practices, etc.) and the health of the soil. The soil therefore is an indicator that can be measured to determine whether a particular land-use activity will negatively impact the delivery of ecosystem services. Soils are also an early warning indicator that management may need to change its practices since these activities may push soil physical, biological or chemical processes towards a tipping point. Those elements of soils that are especially critical in maintaining plant productivities are most frequently altered by human land-use activities, e.g., soil carbon stocks and nutrient storage and delivery. When soil carbon stocks decrease, this carbon is released as carbon dioxide and potentially contributes to climate change. Since the 19th and 20th centuries, agricultural fields have lost 60% of their carbon storage capacity and a quarter of the soils are less healthy and less able to produce crop yields that were possible in the past. The Sumerians started the process of degrading their lands due to agricultural practices that resulted in a 40% decrease in crop yields, which ultimately contributed to their collapse since they could not feed their people [4]. We need to make sure that we can detect changes in soils due to the impacts of our land-use activities and to ensure that our activities do not increase societal vulnerability to decreasing soil health and productivity. This is why books on standardized soil and plant analysis methods are imperative.

Part II
Introductory Methods in Soil and Plant Analyses

Before presenting the specifics of soil physical, chemical and biological methods, it is important to briefly discuss soil and plant sampling protocols and sample preparation steps. If the appropriate sampling design or methods are not implemented, the credibility of the results may be questioned. It may also be difficult for another researcher to duplicate the results. It is also important that any sampling is representative of the site and is not unduly influenced by some unique aspect of local variation. The purpose of soil subsampling (see Appendix A) is to understand soil characteristics over larger land areas in an efficient manner and at the usual scale of management or at the scale where the impacts of land-use can be detected.

One also has to briefly present appropriate laboratory practices since it is important that soil or plant samples are not contaminated during the analytical process (Appendix B). Therefore it is imperative that a researcher is able to maintain uncontaminated conditions and reduce the risk of the containers contributing to a contamination problem. The results are only as good as the researcher's procedural ability. Furthermore some of the chemical analyses are attempting to detect micro-level changes in some of the trace elements. So any contamination introduced by even the flasks or counter tops could mask any chemical changes requiring the researcher to obtain new samples and reanalyze them, if the contamination is even recognized.

Chapter 2

Field Characterization of Soils to Establish Sampling Protocols

2.1 Soil Sampling Design and Methods

If the analytical data you present at the end of a project is to correspond to the real world (or that part of it represented by your field site), you must setup your sampling design carefully. Taking into account the questions you are trying to answer and the resources available to you (time, money, special equipment, etc.), your sampling design will tell you how to collect many samples and where to collect them.

Soils are naturally heterogeneous; their properties vary horizontally across the landscape and vertically down the profile (spatial variability), as well as through time (temporal variability). Spatial variation may be significant at distances of a meter or even a few centimeters, depending on the parameter being measured. The objective of sampling is to obtain reliable information about a large body of soil (the **population**) from a relatively small subset (the **sample**). If the sample is not collected properly, it may not be representative of the population, and therefore the information obtained from it may not be reliable.

2.1.1 Introduction to Sampling Design

The soil population to be sampled should be subdivided, both horizontally and vertically, into sampling strata that are as homogeneous as possible. Several sources of variation within the population should be sampled if valid inferences are to be made about the population from the sample.

Several different sampling designs are available for selecting which **sample units** (soil cores, spadefuls of soil, plants, etc.) to include in a sample [25–27]. In a **simple random sample** each sampling unit of the population has an equal probability of being selected. Random number tables are used to ensure that selection is truly random (unbiased)—although some would argue that the tables themselves are not **truly** random. If the population is broken into a number of subpopulations (for example, a watershed may consist of three distinct elevational forest types), and then simple random sampling is conducted within each, some of the variation (between different subpopulations) is removed from the sampling error (variation among similar sampling units). This procedure is termed **stratified random sampling.** Some previous knowledge of the population is necessary for making proper judgments about stratification.

By superimposing a regular grid on an area and selecting sample units at grid intersections, complete coverage of an area can be assured (which is not always the case with random sampling). Such a **systematic sample** may give increased precision over a random sample as long as the sampling distance does not coincide with periodic variations in the population, such as furrows in a field or regularly spaced hills and valleys in a large forest.

Prior to developing a sampling design and figuring out how many samples need to be collected at each field site, the variability in soil characteristics needs to be examined in the field. This allows you to determine that you are collecting a sufficient number of samples to characterize the soils at the field site. Soil pits dug in the field can be used to characterize and determine what a typical soil profile is for a site. There are several tests that can be conducted in the field to collect this data. These tests provide important information on a field site where no prior data exists and suggest which chemical, physical or biological tests should be pursued in the laboratory. These tests provide clues as to a soils acidity, whether the soil contains carbonates, how aerobic or anaerobic the soil is, and how much biological activity is occurring in the soil due to soil animals or root growth to name a few.

2.1.1.1 Accuracy, Bias, and Precision

Several terms related to sample measurements should be defined and are presented in Table 2.1.

Sources of errors

Errors are present in all samples and measurements. Use of the term "error" doesn't imply that someone has made a mistake; it simply means that an individual observation doesn't coincide with the "true" value of the parameter under consideration. Errors arising from soil sampling fall into four general categories [25]:

Tab. 2.1 Definitions of some sample measurement terms [26].

Accuracy	The degree of agreement of a measured value with the true or expected value of the quantity of concern
Bias	A systematic error inherent in a method or caused by some artifact or idiosyncrasy of the measurement system. Temperature effects and extraction inefficiencies are examples of the first kind. Blanks, contamination, mechanical losses, and calibration errors are examples of the latter kinds. Bias may be either positive or negative, and several kinds can exist concurrently so that net bias is all that can be evaluated, except under special conditions
Precision	The degree to which repeated measurements on a sample agree with each other

1. **Sampling error** is caused by the inherent variation among units of the population. The sample includes only the selected sampling units rather than the entire population. Sampling error can be avoided completely only by including all of the population in the sample.
2. **Selection error** occurs when some units of the population are selected with a greater or lesser probability than was intended, e.g., a tendency to avoid rocky sites in a field or to over-sample the borders of a field, or failing to thoroughly mix soil in a bag before collecting a subsample.
3. **Measurement error** is that error caused by the failure of the observed measurement to be the true value for the unit. It includes both the random errors of measurement, which tend to cancel out with increased sample size, and biases, which are usually independent of sample size. Examples of random errors are those caused by assuming constant weights for cores of soil which have variable weights and those caused by chance variations of technique in the analytical procedure. Biases, independent of sample size, would result if tare weights or blanks were ignored or if a calibration curve were offset to one side of the appropriate curve. If a biased technique is used, such as sub-sampling from the top of a container of improperly mixed soil (selection error) or using an analytical test which gives a reading too high (measurement error), then this error, or bias, would not be included in the computed sampling error. Only constant attention to technique can hold these biases to a minimum, and even then no estimate of their magnitude is ordinarily available.
4. **Calculation error** can occur once an analytical result has been obtained. In most cases, laboratory results must be converted to another form to be meaningful to the investigator. For example, milligrams of calcium per liter of soil extract might be converted to kilograms of calcium per hectare in the zero to 15-cm layer. All conversion factors, dilutions, blank subtractions, etc. must be included in the calculation.

2.1.1.2 General Considerations on Soil Sampling

Soil samples are obtained to allow qualitative or quantitative determination of soil properties that cannot be adequately estimated by field observation.

The first step in sampling is to define clearly the purpose for which the samples are to be used. The properties investigated generally fall into three major categories—physical, chemical, and micro-biological—often requiring different types of samples and methods of sampling. Whether or not results are to be expressed on an area or volume basis must be considered. The second step, then, is to choose the kinds of samples to be taken and the sampling tools to be used in order to fulfill the objectives as effectively and efficiently as possible. The physical characteristics of the soil, particularly stoniness, may limit the choice of sampling techniques and tools. The third step is to determine the number of samples required. This is not always a simple decision and it is generally better to do the job too well if there is doubt about effectiveness. It is often difficult to determine how many samples are needed, since the number depends on the variation in soil horizons and the variation in samples, both of which may be unknown.

2.1.1.3 Common Sampling Tools and Techniques

Types of sampling tools

1. **Core samples**: These are taken in a ring or cylinder driven into the soil by hammering or other means. The sample is retained in a relatively undisturbed condition so that it is suitable for measurement of properties involving pore space and other physical characteristics, particularly field capacity, total pore volume, aeration, etc. These samples are also used for bulk density (mass per unit volume) which is necessary to convert analyses which are made on a mass basis, such as chemical tests, to a volume or area basis for the original natural soil.

2. **Clod samples**: These are natural chunks of soil which are sometimes used instead of core samples when even less disturbance is desired. These may be coated with wax or plastic and used for such purposes as bulk density determination.

3. **Bulk, loose samples**: These samples are usually air-dried, sieved to remove coarse gravel (> 2 mm), well-mixed and then used for chemical analyses; certain physical analyses which are not dependent on macropores such as particle size distribution; and for micro-biological tests. Drying or storage of samples may influence some chemical and microbiological tests. If

fresh samples are required, storage in air-tight containers or refrigeration is necessary.

Types of samples

1. **Core samples** (from 100 to 500 cm^3 volume): (1) Drive-type samplers with metal or plastic core-sample retaining cylinders; and (2) Auger-flight samplers are similar except that the auger bit principle is used to force the sampler into the soil.
2. **Loose samples**: These may be taken with a spade or trowel by digging a hole or from the surface using a soil auger, soil tube, or bucket auger. From a pint to a quart of soil is usually sufficient for all analyses.

2.1.1.4   Soil Sample Preparation

If samples could be analyzed as soon as and in the same form in which they were collected, there would be no need to worry about sample transport, storage, and preparation. However, except in the case of some simple, usually qualitative field measurements (e.g., sample color), soil analysis occurs in a laboratory separated in time and space from the sampling point.

It is essential to select sample preparation and storage procedures that are compatible with the sample analyses you want to perform. This decision should be made before you begin sampling. Incorrect preparation may alter the analytes of interest, invalidating your later laboratory work.

Field storage

Zipper-type plastic bags are the most convenient way to store samples at the point of collection (remember to squeeze the air out as you zip up the bag, so the sample will take up the least amount of room).

In most cases, it is desirable to minimize any biological activity that might occur in a sample after it has been collected. The mere act of digging up some soil and dumping it in a bag is sufficient disturbance to greatly stimulate rates of microbial activity. Therefore, it is best to refrigerate samples as soon as possible after collecting (using ice, if necessary); or, to begin drying them, if you have decided on that method of sample preparation.

Drying of soil samples

Effects of drying on soil chemical and physical properties have been studied extensively [28, 29]. In general, drying and subsequent rewetting cause problems in the following ways: (1) Organic components of microbial cells are released when the cells are killed by drying; on rewetting, there is a flush of microbial activity, which may further transform the chemistry of the soil solution; this effect is most pronounced with biologically active analytes, such as nitrate, ammonium, and phosphate; total C and N are usually unaffected; (2) Certain clays may irreversibly fix potassium and ammonium ions when dried; some interior cation exchange sites also become unavailable.

Both of these effects are more pronounced as the temperature of drying increases. From a purely physical point of view, clayey soils can be very difficult to work with when dried.

Most soil labs recommend air drying all samples (preferably in a forced-air oven at 30–40°C). For clayey soils, if some amount of continuing biological activity can be tolerated, use of field-moist samples that are stored at 4°C may be preferable.

Another way of looking at the problem is that analyzing a field-moist sample gives you a snapshot at an instant in time, under a particular set of environmental conditions (moisture, temperature, etc.) that can change rapidly. Drying samples (that were probably collected over several days or weeks) before analysis may yield results that vary less rapidly with changing field conditions—in other words, drying is part of **standardizing** a procedure.

Storage Temperature

For dried samples, room temperature (15–25°C) is adequate for sample storage. Considerable microbial activity can occur in field moist soil even at 4°C, however. Freezing will stop this activity, but freeze-thaw cycles have similar effects to drying-wetting cycles—some cells are killed, releasing their contents into the soil solution and providing a substrate for a flush of microbial activity on its thawing.

Sample Transport

Soil and plant samples collected outside the continental United States are governed by permits issued by the U.S. Department of Agriculture Plant Protection and Quarantine Program. The permits specify packing, labeling, storage, and disposal requirements. Briefly, samples should be packed in "sturdy, leakproof containers" (double-bagged in a cooler or cardboard carton is good); the outer container should be labeled as "Soil [Plant] samples", have a copy of the soil or plant import permit, and be prepared for an inspection.

In the laboratory, sample labels should include the country of origin. Do not keep samples longer than necessary, and autoclave samples before throwing them away.

United States Department of Agriculture (USDA) inspects laboratories in the U.S. that conduct soil and plant analyses every year (if the laboratory resides in some other country there is probably an equivalent inspection agency and program there). Please cooperate with this program to control the spread of agricultural pests and/or diseases.

Check with your airline before using dry ice to cool samples that are to be shipped by air.

Paper bags are good for air-drying samples, as long as they're not too wet to begin with. Prolonged oven-drying may make the bags brittle, though.

For a medium- to long-term storage, zipper-type plastic bags (with the air squeezed out) are ideal, although there is some gas exchange through the plastic that may affect samples over very long times. Glass or plastic jars with tightly sealing lids are also suitable.

"Soil" (or "fine-earth fraction") is conventionally defined as the less-than-2-mm fraction of whatever you dig out of the ground and put in a bag. The coarse material (gravel, rocks, and coarse roots) that doesn't pass through the 2-mm sieve is much less active, chemically and physically, due to its low surface-to-volume ratio (see Section 5.3, Soil Texture). Virtually all soil analysis is performed on the fine earth fraction, so soil samples are routinely sieved. Silt, clay and organic matter tend to make soil sticky when wet, so it is usually easier to sieve dry soil. Some clayey soils, however, become very hard when oven-dried, or even air-dried, so experience will have to guide you.

Sieved and dried soil samples often undergo a "self-sorting" process in which slight movements of the sample bag or container cause the finer particles to accumulate at the bottom. Before you take a subsample of the whole for further processing or analysis, thoroughly mix the sample to distribute all grain sizes uniformly.

For analytical techniques that require very small amounts of sample, it may be necessary to grind the soil so that the maximum grain size is much less than two millimeters. Jackson [30] suggests that a maximum particle size be selected such that the sample required contains at least 1,000 particles of that size. For example, a 13-mg sample should be ground to pass a 60-mesh (0.25 mm) sieve.

2.1.2   Soil Sample Process Procedure [1]

1. Using the Munsell charts, note color of soil. Give complete designation (Hue, Value, and Chroma). Record as Moist Color.
2. Note any other characteristics of your sample as you work with it: structure, consistence, uniformity, etc. Also, record any other information giving you about where and how the soil was sampled.
3. Weigh an empty plastic bag. Record as "Tare weight".

---

1 See Appendix C for examples of data sheets.

4. Cover the bench top with a large sheet of brown paper. Use a smaller piece of brown paper to collect the soil from the sieve.
5. Pass your field-moist soil sample through a 2-mm sieve and pour it into the plastic bag for storage.
6. Weigh the sieved soil. Record as "Tare+Soil". Calculate Soil weight by subtraction.
7. Weigh rocks and roots remaining in the sieve. Calculate each as a percent of total soil weight. See Equation (2.1) as a calculation example,

$$\text{Percent Rock} = \text{wt.rocks}/(\text{wt.soil} + \text{wt.rock} + \text{wt.roots}) \times 100\% \qquad (2.1)$$

Note that these are "wet" weights; they can be corrected later for water content.

2.1.2.1 Sample Variability—Number of Samples Required

The main questions here are: first, what is soil variability? second, how do we measure or estimate it? and third, how do we take it into account in sampling? Soils are characteristically variable, and forest soils are more variable than agricultural soils. That is, they are not uniform from point to point because of microtopography or other factors.

Variability is quantified by two statistics (for more discussion on this see Appendix A): the variance, s^2, and the standard deviation, s (the square root of the variance) as shown in Equation (2.2) [26]. The variance is an average of the squared deviations of each individual sample observation from the mean (squaring makes positive and negative deviations equivalent):

$$s^2 = \frac{\sum_i (x_i - \bar{x})^2}{n - 1} \qquad (2.2)$$

$$s = \sqrt{s^2}$$

where x_i=an observation, $\bar{x}$=sample mean, and n=the total number of observations,

There is a simpler calculating formula for s^2 and s, which doesn't require you to know the mean beforehand:

The variability associated with particular soils, horizons, and sampling methods has to be learned through sampling and analysis until some basis of judgment is built up. As an example, in a study of humus layers and their moisture properties, 15-cm-square areas were sampled, taking four samples each week from each area to estimate moisture content.

The questions to ask are: "How good is the average value and how many should we take?" The four samples in the first week gave values of 26.8%, 57.7%, 37.2%, and 32.6% in the Oe layer (a surficial organic layer according to the U.S. "Keys to Soil Taxonomy") [31], or a mean value of 38.6%. The precision of these observations is measured by their standard deviation, s, which is the square root of the variance, s^2. See the following calculations (Tab. 2.2).

Tab. 2.2 An example of calculating the variability and precision associated with sampling.

i	x_i	x_i^2	s
1	26.8	718.24	
2	57.7	3329.29	
3	37.2	1383.84	
4	32.6	1062.76	
	$(\Sigma x_i) = 154.3$	$\Sigma(x_i^2) = 6494.13$	
	$(\Sigma x_i)^2 = 23808.49$		13.5%

where

$$s = \sqrt{\dfrac{6494.13 - \dfrac{23808.49}{4}}{3}} = 13.5(\%)$$

i.e., 68% of the observations should fall within the range 38.6%±13.5%, or 38.6%±27% would be expected to include 95% of the observations.

The accuracy of the mean is measured by the standard error of the mean, $s_{\bar{x}}$, which is estimated by:

$$s_{\bar{x}} = \sqrt{\dfrac{s^2}{n}} = \sqrt{\dfrac{180.7}{4}} = 6.72$$

i.e., 38.6%±6.72% would be expected to include the true mean with a probability of 68% and 38.6%±13.44% would include the true mean with 95% probability. As you can see, increasing n increases the accuracy of your estimate of the true mean.

The **coefficient of variation** (Eq. (2.3)) is used to obtain a relative value of the degree of variability, by expressing the standard deviation as a percent of the mean (CV is also called %RSD, or relative standard deviation):

$$CV = \dfrac{s}{\bar{x}} = \dfrac{13.5}{38.6} = 35\% \tag{2.3}$$

The smaller the percent, of course, the less variable the observations are.

From these statistics you can estimate the number of samples needed for the accuracy desired, or the accuracy to be expected from a certain number of samples. For instance, if 10 samples were taken:

$$s_{\bar{x}} = \pm\sqrt{\frac{180.7}{10}} = \pm 4.25$$

or to reduce the standard error of the mean to $\pm 2\%$ moisture content:

$$s_{\bar{x}} = \pm\sqrt{\frac{180.7}{n}} = \pm 2.0$$

$$n = \frac{180.7}{(2.0)^2} = 45 \text{ samples}$$

One of the most common ways to overcome the problem of horizontal variability when sampling is to take a large number of small samples in a random pattern and physically mix them together to obtain an "average" material. This is called **sample compositing**. While compositing may result in savings of analysis time (and therefore money), less information is gained on soil variability. If sampling costs are higher than analysis costs (usually not the case), then there is no cost advantage to compositing.

Variability in the vertical direction is generally associated with quite different horizons. These differ in thickness and in properties from point to point so each horizon may be more or less variable than other horizons. Theoretically it might be desirable to take a different number of samples from each horizon. Compositing of samples from a vertical section through the soil presents some difficult problems because quite different materials are mixed together for analysis. However, shallow fixed-depth samples of about 15 cm are often used.

Another very serious problem is that the variability for different characteristics may not be the same, e.g., a coefficient of variation that is satisfactory for determining water-holding capacity may not be adequate for estimation of calcium content.

The following formula (Eq. (2.4)) can be used to calculate the sample number needed for statistical sampling.

$$n = \frac{t^2 \times s^2}{D^2} \tag{2.4}$$

where n is the number of samples. t is a "t" distribution value is found in the "t" table by choosing a chosen level of precision (in the following example shown, 95% precision is selected) and then a number of degrees of freedom (df) are chosen. Then find the value at the intersection of the column of the precision level and the row of the degrees of freedom. The degrees of freedom for "t" are first chosen arbitrarily (say 10) and then modified by reiteration. s^2 is the variance which is known beforehand from other studies (or it is estimated where

R is the estimated range likely to be encountered in sampling) [32]; and D is the variability in estimation of the mean that you are willing to accept.

For example, for a soil with a polychlorinated biphenyl (PCB) concentration ranging from 0 to 13 ppm [1] , the estimated sample number at 95% probability and within 1.5 ppm of the true mean (confidence limits) is:

$$n = \frac{(2.23)^2 \times (3.25)^2}{(1.5)^2} = 23$$

Since 23 samples are many more than the 10 we used to obtain a "t" value, we must run another iteration using a "t" value equal to our new estimate:

$$n = \frac{(2.069)^2 \times (3.25)^2}{(1.5)^2} = 20$$

Testing 20 samples for PCB is expensive, but the estimated variance is high (a range from 0 to 13 ppm). Any sampling scheme that reduces the variance will lower the number of samples needed. The number can also be lowered by relaxing the probability from 95% to 90% or by allowing the confidence limits to increase to ± 2 ppm or higher.

The ultimate goal is to sample with the precision necessary to satisfy the sampler's objectives. One can cut down on sample numbers if the variability is low, if less precision is needed, or if a higher probability of sampling error is acceptable (say 10% as opposed to 5%). However, in cases where more than one property will be measured on each soil sample, the total number of samples should be based on the property requiring the highest sample number.

Frequently it is desirable to measure certain soil characteristics as they occur in nature over a period of time, particularly moisture and temperature levels. Conventional methods of sampling can be used but repeated sampling results in "no two samples being obtained from the same place". Because frequent heterogeneity exists in soils and their properties, a high variation may be found between samplings that could just be due to differences in the samples analyzed rather than due to temporal changes in the parameters measured. Frequent sampling could also result in high local soil disturbances and eventual destruction of the site. For this reason, devices that can be placed in the soil to record moisture and temperature, etc. continuously or intermittently at the same point without disturbance have been developed. Examples of soil moisture and temperature devices are soil moisture blocks, time domain reflectometry (TDR) probes, and neutron soil-moisture measuring equipment.

---

1 1 ppm$=10^{-6}$.

Chapter 3

Plant Tissue Characterization

Soils may be analyzed for soil nutrients when conducting either basic research (e.g., ecosystem budgets, etc.) and/or for applied reasons (e.g., nutrient pool for plant growth in agriculture or forestry, etc.). Similarly, plant tissues are analyzed for nutrients for the same reasons. Interestingly, plant nutrient analyses may even be analyzed as an indicator of the soil nutrient status. Researchers recognized the links between soils and plant nutrient status more than a century ago. For example, the director of the Rothamsted Experiment Station published a paper in the *Journal of Agricultural Science* entitled "The Analysis of the Soil by Means of the Plant" in 1905. Similar work had in fact been going on since the days of Theodore de Saussure who, in 1804, showed that the composition of plant ash, which reflects the nutrients taken up by a plant, varies with soil type and the age of the plants being analyzed. He also showed that the ash mainly consisted of alkalis and phosphates. He further showed that plants grown from seed in water—no soil growth medium—did not produce ash when combusted which suggested that ash constituents were taken up by plants from the soil [33–35].

Determination of the nutrient fraction that is realistically available to plants from a soil presents the most difficult problem for soil scientists. Regardless of the nature and strength of the extracting solutions, chemical analysis can provide only an approximation of the dissolving ability of root systems, especially since most plant roots have formed symbiotic associations with micro-organisms (see Sections 8.1.1 and 8.3). This discrepancy has led to the idea of utilizing the root systems of living plants as the natural extracting agent and determining available nutrients by analysis of either tissues of entire plants or their foliage. This approach seems to provide a simple approach to the problem of assessing plant nutrient deficiencies and how many fertilizers to apply to plants. However, research has disclosed that plant tissue or foliar analysis can be more complex and interactive than what might initially be thought [36].

Specialists concerning with increasing the production of citrus crops have specifically used foliar analyses to monitor their crops. The results of research on citrus crops indicate that by relying solely on the results of foliar analysis can be misleading regarding how well citrus crops are growing. Other factors, such as injury by parasitic organisms, adverse climatic influences, and composition of the soil, influence the results produced by foliar analyses. Their research suggested that tissue sampling must be confined to a specific tissue age and to foliage with specific characteristics, e.g., 5- to 7-month-old bloom-cycle leaves from non-fruiting terminals. Interpreting results of foliar analyses has to balance each nutrient in relationship to other nutrients that are also detected. This means the researchers should develop a ratio of all nutrients in the leaves. If one nutrient element is unusually deficient in a leaf, it will influence the nutrient contents of other nutrients found in that leaf.

Sampling leaves for their nutrient contents has utility in some situations where the growth conditions are very controlled and soil conditions are relatively uniform. This is in contrast to the situation found for plants growing under natural conditions and where the soil characteristics can vary at the centimeter scale. For example, foliar analyses have been very useful when appraisals are needed for the fertility of nursery soils and when greenhouse studies are assessing the effect of fertilizers and biocidic compounds on greenhouse plants. Foliar analyses are more difficult to use to evaluate soils nutrient levels when plants are growing in farmers' fields or under the less controlled conditions found with forest plantations. Under these situations, the spatial variability in soil types and characteristics at small spatial scales is difficult to sample [37]. This variability in soil characteristics means that it is difficult to assess the future production potential of crops to be planted in fallow lands using foliage samples collected from the previous crop that grew in these fields.

Other factors complicate the use of foliar analyses to determine the fertility of a given piece of land. For example, depending upon the age of a plant, it may be acquiring nutrients from a different part of the soil profile. Young tree plantation roots are frequently confined to the superficial layer which is often depleted by previous agricultural uses of the land. In other cases, nutrients are removed from the foliage by rainfall. Nutrient leaching from leaves is especially common during foliage senescence. Therefore, sampling of foliage must be accomplished at the end of a growing season but before nutrient translocations that preface leaf fall. This limitation of the sampling period and the long time necessary for sampling a single plant reduce the use of foliar analyses at the large scale needed to assess the soil nutrient status of even a 10 m by 10 m plot.

3.1 Tissue Sampling

Analysis of plant tissues may include three broad categories of organic materials:
    1. Live plant tissue,

2. Senesced tissue (and/or partly decomposed) such as forest floor organics, and

3. Soil organic matter (SOM).

The analytical procedures are generally same for all these categories or organic materials. The greatest differences would involve the sampling design and, of course, the methodology needed by each organic material. For example, the SOM needs to be separated from soil mineral particles which is not an issue with live or senesced plant tissues.

Some considerations of sampling designs when analyzing "live plant tissues" include whether to sample:

Whole plant (above-and/or below-ground) parts

The choice of plant part for sampling and analysis depends upon the kind of plant and purpose of the assay. Whole plant samples of small plants are easy to take and analyze. However, such a sample often includes tissue differing in age and function, and the analysis value represents an average. This value also depends upon the rate of plant growth and proportion of older tissue to young tissue, as well as the nitrogen (N) supply in the soil. Values from the analysis can be interpreted if levels of deficiency and adequacy are known for that particular plant at the same stage of growth. However, sampling the entire plant will likely tend to mask important differences at key sites within the plant. Whole plant samples are necessary in N research when total uptake is part of a study of fertilizer-use efficiency or N balance. Older and larger plants are more difficult to sample, dry, and grind for analysis. Tree crops are certainly the least suitable for whole plant analysis.

Leaf and stem tissues

Leaf samples, often including the petiole, are most commonly used for Kjeldahl N determinations, either with or without the inclusion of NO_3^-. Again, these values better indicate the cumulative N status until the time of sampling and are not necessarily an accurate representation of current status. Nitrate transported to the leaf blade is rapidly reduced in most plants with normal growth. Stem tissues contain much higher concentrations of NO_3^- than leaf tissues; NO_3^- is also stored in associated parenchyma cells. For example, with an adequate N supply, a cotton petiole may contain 20 or more times as much NO_3^- as the leaf blade. Furthermore, the change in NO_3^- in the petiole is much more sensitive to that in NO_3^- supply to the plant. Stem tissue thus appears the most suitable plant part to sample and analyze for evaluation of current N status of actively growing annual crops. However in woody-stem perennial plants (e.g., trees), leaf tissue generally has been found to correlate well with N status.

Position on plant for sampling

Nitrogen compounds in plants are continuously degraded, retranslocated, and resynthesized into new compounds. The older leaves are poor competitors for receiving the breakdown products because of internal competition for N in a plant. New leaves are stronger competitors and fruits are even stronger sinks for metabolites. Thus, leaf age is a more important consideration than leaf location. On annual plants, the lower leaves are commonly the older leaves, but on trees, the old and new leaves can occur over the entire plant.

What are most important in the diagnosis of the N status of a plant is determining what plant part and where to collect samples for each plant. The protocols need to be standardized so that the results can be interpreted correctly. To determine the current plant N status, a soluble N analysis should use petioles or stem tissues collected from recently and fully developed leaves or main stem tissues that are located above the soil level. For trees it is not uncommon to analyze the current and/or mature foliage (depending upon the question asked) with some standardization of sampling zones to include the top, middle and bottom 1/3 of the canopy (e.g., shade leaves versus sun leaves). This sampling has been further standardized by collecting tissue samples growing on the south side of the trees.

Stage of growth

In theory, the concept of critical nutrient concentration is useful for separating the zone of adequacy from the zone of deficiency in terms of nutrient concentration within the plant. However, it fails to serve as a practical guide to measure the N fertilization impacts on growing crops/plants throughout the season. The NO_3^- level detected in a plant tissue sample must be interpreted differently at different physiological ages of the plant. For example, different plant parts or tissues may accumulate or store some nutrients at greater or lower contents than other parts or tissues at different times of the season. Even the nutrient concentrations may appear to change (within the same tissue) seasonally. This may just be an artifact of the growth dilution. Some researchers have attempted to account for this by expressing the leaf nutrients on a "per 100 needles" or "per one leaf" basis instead of a "per needle" or "per leaf mass" basis. When using these values as a guide to determine how much fertilizer to apply, the decline in NO_3^- to a deficient level must be anticipated by its rate of decline as growth is stimulated by the fertilizer applications. Fertilizer can then be applied before the plant NO_3^- closely approaches this critical value. Periodic sampling and analysis of the plant during the season can serve as a postmortem evaluation.

Such use of this diagnostic technique is valuable in assessing the N soil fertility program, and, to some extent, in planning for the next year's crop. Values from samples at late stages of growth usually fall into this category, even if earlier values are used to guide N fertilization. Some research [38] suggests that nitrogen and potassium concentrations decrease with maturity of the plant, phosphorous changes relatively little, and calcium and magnesium tend to increase with the maturity of the plant.

Sampling designs for the senesced tissue

– Tissues recently senesced but still standing or attached to a live plant can be obtained from the plant in a similar manner as described for "live plant tissue" as previously discussed.
– Tissues recently senesced but on the ground can be obtained by random or stratified-random sampling and even separating by species and tissue types, e.g., leaves, fine woody debris (<1 cm diameter), coarse woody debris (>10 cm diameter), flowers, seeds, etc. These samples are part of the Oi horizon.
– Tissues that have been fragmented and/or decomposed can be separated into the Oe and/or Oa horizons and may also be randomly or stratified-randomly sampled.

Sampling designed for the SOM may also be randomly or stratified-randomly sampled with a soil corer or similar device. The SOM then has to be separated from the mineral soil. Some methods would include density fractionation using NaI or Na metatungstate [39] and then analyzing the fractionated organics.

3.2   Tissue Preparation and Laboratory Extraction

Preparation of the obtained plant tissue involves several steps. If surface contamination would influence the analysis (e.g., dust containing Fe when Fe is to be analyzed), then a light washing may be necessary (too much washing may also leach out some nutrients, e.g., K). The collected tissue should be dried as soon as possible in a forced-air oven. The temperatures should be maintained between 60°C and 80°C; at higher temperatures some nitrogen fractions might be lost through volatilization. After the material is dried, it is ground in a Wiley mill to pass at least a 20 mesh sieve. The ground material is thoroughly mixed and stored in air-tight jars. Some investigators recommend the use of plastic gloves when handling plant material, glassware and crucibles in the laboratory.

In the determination of trace elements it may be necessary to grind the tissue in a mortar by hand, a procedure which minimize contamination of samples. Contamination is also considerably reduced by the use of a Wiley mill with a special stainless steel screen.

Because results of analyses are expressed on the "oven-dry" basis, it is necessary to determine moisture content of "air-dried" tissue samples if the analysis is actually conducted on "air-dry" tissue. Of course "air-dried" tissues generally have significantly more water in their tissues than "oven-dried" tissues. However, many analyses are conducted on "air-dried" tissues because in the act of "oven-drying" some critical nutrients may be volatized or even complexed thereby decreasing their final analytical concentrations. Results of air-dried tissue analyses are later converted to the oven-dried basis. Duplicate determinations should be made on each sample of plant tissues. If any tissue is contaminated by soil minerals, it will require a separate oven-dried (60–80°C) subsample to be ashed (or combusted) at 500°C overnight to determine the ash percentage. The weight of the sample analyzed may then be corrected to determine organic weight only (or ash-free weight).

Tissue samples can be analyzed for nutrients using the Kjeldahl method and also the dry-ashing with HNO_3 method (see Chapter 7). Plant nutrients, however, may be analyzed by methods other than just by the wet or dry oxidation methods previously listed. For example, plant tissues can be extracted just as soil samples. Some common extraction methods are: soluble P and K in acetic acid [40], nitrate by the phenoldisulfonic acid method [40], and mineral elements extracted by TCA (trichloroacetic acid) [41]. The quantity of extracted nutrients are, of course, related to the type of extracting solution used and the form in which the nutrient resides or how it may be bound in the plant tissue.

Chapter 4

Introduction: Laboratory Practices

Please read the following information on safety, cleanup procedures, and general laboratory protocols for soil and plant sample preparation for physical, chemical or biological analysis. You need to adhere to these guidelines, in order to:

- Ensure your safety and the safety of other laboratory users,
- Maintain the equipment and supplies in good working order,
- Maintain a known level of quality control for your experiments and those of others,
- Facilitate a good and cooperative working environment.

Please read this information carefully and keep it for future reference. It is important to remember that "rules and regulations" differ from one laboratory to another depending on the type of research being done. Therefore, it is better to <u>ask first</u> if you have any questions, to be sure that the procedures you are using are appropriate for the laboratory and the work being done—you will save yourself and others a lot of time.

4.1 General Laboratory Protocol

Be familiar with a procedure before you start doing it. Know why you are adding a particular reagent and the likely reaction that will occur. This will reduce the chance of a time-consuming or potentially hazardous mistake.

Be sure that all containers you are using are labeled with the contents and dates if they are to be stored for more than a few minutes. The United States Occupational Safety and Health Administration (OSHA) and Environmental Protection Agency (EPA) regulations require labels with chemical names written out, rather than just formulas or abbreviations—**even "Water" must be written out, not "DW" or "H_2O"**. Take these regulations seriously—a University laboratory can be financially fined if an inspection turns up improperly labeled containers of either fresh reagents or hazardous waste. If you find containers (e.g., squirt bottles) with no label, they are probably filled with distilled water—**but don't assume anything!** Find out first.

If you do not know whether a particular chemical is toxic, hazardous or reactive, look it up in the *Merck Index* (copies should be available in an Analytical Laboratory). That book lists most known, or commonly used, chemical compounds and gives general information on their uses, reactions, interactions, and hazards.

4.1.1 Safety

In many laboratory procedures you will be working with hazardous chemicals and equipment. Safety issues specific to a particular procedure will be discussed during your training for that procedure. If you have a question concerning a potential hazard, **ask first**. Be familiar with the location and use of safety equipment and procedures. As a general rule, it is best not to work alone in the laboratory, but solitude may be unavoidable if you work during evenings or weekends.

Some general safety rules:

1. Wear goggles, gloves, and protective apron or laboratory coat when using any strong acid or base solutions, or strong oxidants.

2. Sandals are **not** appropriate laboratory shoes; closed-toe shoes are best. Long pants are preferable but not required.

3. Pour and mix strong acids and bases under a hood with the exhaust fan turned on. Leave the solution in the hood until you are sure that all fumes have subsided. **never** remove anything from a hood that is not yours without asking first. Never leave a potentially hazardous chemical where it could fall, spill, or get knocked off a counter.

4. All chemical containers must be labeled with their contents (including distilled water), even if they are only being used temporarily. Tape and markers should be available in any laboratory.

5. If you spill an acid, clean it up with lots of water and baking soda (sodium bicarbonate) to neutralize the acid. If you spill a base, clean it up with lots of water and a dilute acid. For major spills, use the spill kits that are typically located by the front door of the laboratory.

6. Use the acid burn or base burn solutions if you get strong acid or base on your skin. Follow with lots of water.

7. Put broken glass and other sharps in the appropriate containers. Otherwise, they are hazardous to anyone emptying trash containers.

8. Know the location of the eye wash stations and safety showers and learn how to use them properly. There should be a first aid kit by the front door of every laboratory where chemicals are used.

4.1.2   Laboratory Water

Three kinds of water are commonly used in the laboratory. For trace-level analysis (part-per-million or less), it is important to use the right kind:

– **Tap water:** This water is full of various things that we usually do not want to be present in laboratory analyses. You can use tap water for the first rinse of glassware. There are some procedures where small amounts of dissolved salts are irrelevant or beneficial, such as determining soil texture with the Bouyoucos hydrometer, or determining soil water potential with the pressure plate.

– **Distilled water (DW):** This water is distilled water that has been distilled once. This water can be used for rinsing glassware. It probably contains low concentrations of dissolved ions that could potentially interfere with some chemical analyses.

– **Nanopure or distilled-deionized water (DDI):** This is distilled water which has been run through de-ionizing resins in a Nanopure system. Use this for making reagents, standards, extracting solutions, blanks, etc. Typically this water must be autoclaved when used to conduct microbiological work, since it typically does not receive a final filtering through a 0.45-μm filter.

4.1.3   Clean-up

Laboratory clean-up and maintenance of supplies and equipment are part of every research project. If you do laboratory experiments, you will create messes that need to be cleaned up. You are responsible for the clean-up of glassware, laboratory benches, and equipment that you use both individually and as a group. Other people are conducting research or class projects in a laboratory and require a clean laboratory environment, just as you do. At the end of each laboratory session check to be sure that all glassware, equipment, and supplies are properly cleaned and put away, and that all counter surfaces are clean. This will not only create a better, safer working environment, but is common courtesy to the rest of those who depend on the laboratory space and equipment for work.

Each laboratory procedure has its own clean-up protocol depending on the particular requirements of the research. Analyses of soil and plant material tend to be messier than straight chemical analysis and require thorough cleaning. Be sure to include time for cleanup when scheduling lab work.

The basic clean-up procedure for ceramic-, glass-, and plasticware

1. If the item has come in contact with soil, soil solution, or plant material, wash with tap water and phosphate-free detergent to remove any residue (visible or microscopic).
2. Rinse well with tap water to remove detergents.
3. Rinse with a small amount of dilute acid (1 mol L^{-1} HCl). The acid bath may also be used, for large numbers or small size items.
4. Rinse 3 or 4 times with distilled water (DW).
5. Place inverted on a clean paper towel to dry, or on a drying rack. Do not use wooden peg racks located over the sink–they cannot be kept clean enough for trace-level analytical work.
6. Come back in a day or two to put away the dry labware.

The basic clean-up procedure for metal

1. Metal objects (except for soil cans; see Section 4.1.3.3) are washed with water (and detergent if necessary but **no acid**), then distilled water.

The basic procedure for the care and use of soil cans

1. Some soil cans have sharp inside edges. Be very careful when handling and cleaning them.
2. Label all cans with a unique identification (ID) number for recording purposes for all future analytical work. Always store each can with its own lid. They should both have the same ID number.
3. Put cans away in their proper location (by size). Metal soil cans are typically cleaned, sorted by size, and labeled. Many people use them, so please return them as soon as possible.
4. Don't put tape on cans. Don't put any additional identification numbers or unnecessarily mark the cans. They are already marked with an ID number. Use the ID number already on each can and lid. Then you won't have to put any additional marks or tape on them.
5. Don't forget to fill in the information sheet on the oven door indicating what samples you are placing in the oven and your name. Don't leave your samples in the oven any longer than necessary—oven space is always at a premium in any laboratory.
6. Clean out your cans as soon as possible, so they can be used by the next person.

4.1.4   Waste Disposal

Used filter paper, weighing boats, paper towels, etc. are disposed of by putting them in the regular trash cans. **Glass and sharps (including pipettor tips) go in special sharps containers.**

Hazardous Wastes Disposal: Hazardous chemicals must be properly disposed of. If you don't know if something is hazardous, please ask; think before dumping anything down the sink. Before beginning a new procedure, find out how to dispose of the waste, residues, etc. **Hazardous waste containers must be properly labeled, with name(s) (spelled out, not chemical formula or abbreviation), approximate amount or concentration, and the words "HAZARDOUS WASTE".** There are important legal as well as personal safety considerations involved, so please follow the rules!

Disposing of Soil: Do not put soil down the sink!!! Use soil traps to get rid of soil suspensions. Putting soil down the drain can clog up the whole building and may require ripping up the basement floors.

Non-local soil must be autoclaved before disposal (to prevent the spread of nematodes and other peripatetic pests), so it goes into separate containers from local soil. This is a US Department of Agriculture requirement (outside of the U.S. check requirements for your own location). The autoclave bucket should be conveniently located on the left side of the sink. Do not fill it more than half-full, or the bag won't fit in the autoclave. Try to minimize the amount of liquid that goes in the bucket.

Please take responsibility for cleaning out the soil trap before it gets full:
–   After sediment has settled, drain off the water.
–   Dump sediment into trash (domestic soil only [1]). Use one of the large trash cans with heavy-duty bags, **not** the small waste basket that might be located next to the sink.

---

1 Remember, non-local soil goes into the autoclave bucket for sterilizing in the autoclave before disposal.

Part III

Soil Physical, Chemical and Biological Analyses

Chapter 5

Methods for Analyzing Soil Physical Characteristics

A soils physical characteristic strongly influences the interconnectivities between plants and soils. Human land-use activities impact soil physical characteristics by compaction and decreasing soil organic matter contents. These changes all decrease plant held water, the erosion potential of a soil and also its nutrient delivery capacity. Managing soils on a long-term basis for agricultural production has a significant impact on soil physical characteristics because of the loss of organic matter from soils.

Soil compaction that occurs during land-use activities also decreases a soil's porosity—the amount of air or water filled space between soil particles—which decreases water movement into a soil and can cause surface runoff and erosion to increase. Therefore the physical characteristics of soils determine the rate at which water moves the soil and how much water will be held for plants. If a site has fine textured soils, this soil will hold more water than a coarse textured or sandy soil.

Soil physical characteristics are important to measure since they will determine a soils water holding capacity, its structure, its aeration, and how much nutrients will be stored on soil particles. Soil structure—how soil particles form larger clusters or peds—is part of the physical characteristic of a soil. This structure determines a soils erosion risk, how water moves or does not move through a soil, and whether a soil will be fertile or sequester carbon as organic matter.

Soil physical characteristics typically measured are soil bulk density, soil texture, soil moisture contents and soil water potential. These will be briefly introduced next.

5.1 Soil Moisture

The amount of water associated with a given volume or mass of soil ("soil water" or "soil moisture") is a highly variable property. It can change on time scales of minutes to years. However, most soil properties are more stable, and should be referenced to dry soil weight. The example below shows why a dry soil weight reference and not a wet weight reference should be used in soil analyses.

Example: A 100-gram sample of moist soil is found to contain 1 gram of nitrogen, for an N concentration of 1%. On drying, the sample weighs 67 grams (i.e., one-third of the moist sample was water). Nitrogen now is 1.5% of the sample. On standing exposed to humid air, the sample absorbs 8 grams of water, and N concentration now "drops" to 1.33%. If the soil water concentration at the time of analysis is determined, N concentration or any other property can always be expressed relative to dry soil weight, regardless of how the soil water changes with time.

Of course, water concentration itself is often a parameter of interest, particularly in field studies of plant water relations, hydrology and stream flow, or crop production and irrigation.

Expressing Soil Water Concentration: There are several ways to express soil water concentration (Eq. (5.1)–(5.11)) (Note: the Greek letter theta, θ, is also commonly used for water concentration):

Gravimetric—dry-weight basis

$$w_d = \frac{\text{grams of water}}{\text{grams of dry soil}} \tag{5.1}$$

Range: 0 to infinity. This is the form that is generally accepted and used.

Gravimetric—wet-weight basis

$$w_m = \frac{\text{grams of water}}{\text{grams of moist soil}} \tag{5.2}$$
$$= \frac{\text{grams of water}}{\text{grams of water} + \text{grams of dry soil}}$$

Range: 0 to 1 (0 to 100%).

To convert between dry and wet basis:

$$w_d = \frac{w_m}{(1 - w_m)} \tag{5.3}$$

$$w_m = \frac{w_d}{(1 + w_d)} \tag{5.4}$$

Volumetric

$$w_v = \frac{\text{volume of water}}{\text{volume of soil}} \tag{5.5}$$

Range: 0 to 1 (0 to 100%).

To convert:

$$w_v = w_d \times \frac{\rho_b}{\rho_w} \tag{5.6}$$

where ρ_b=soil bulk density (g dry soil/cm^3 soil), ρ_w=density of water (1 g water/cm^3 water).

Depth

$$w_z = \text{cm of water in a given depth zone of soil}$$

To convert:

$$w_z = w_v \times \text{depth} \tag{5.7}$$

For example, how much water is in the top 15 cm of soil that has a volumetric water concentration of 0.2 cm^3/cm^3?

$$w_z = 0.2 \text{ cm}^3 \text{ water/cm}^3 \text{ soil} \times 15 \text{ cm soil}=3 \text{ cm}^3 \text{ water/cm}^2 \text{ soil}$$

$$= 3 \text{ cm water depth per unit area of soil}$$

Calculating Gravimetric (dry) Soil Water Concentration:

$$w_d = \frac{(\text{Moist soil wt.} - \text{dry soil wt.})}{\text{Dry soil wt.}} \tag{5.8}$$

Calculating Gravimetric (wet) Soil Water Concentration:

$$w_m = \frac{(\text{Moist soil wt.} - \text{dry soil wt.})}{\text{Moist soil wt.}} \tag{5.9}$$

Calculating dry weight equivalent of a moist soil sample:

$$\text{Dry soil wt.} = \frac{\text{Moist soil wt.}}{1 + w_d} \tag{5.10}$$

or

$$\text{Dry soil wt.} = \text{Moist soil wt.} \times (1 - w_m)$$
$$= \text{Moist soil wt.} \times (100\% - w_m\%)/100\% \tag{5.11}$$

Note: w_m is the "water fraction" of a moist soil, so $(1 - w_m)$ or $(100\% - w_m\%)$ is what's left, which could be called the "dry soil fraction".

Calculating how much moist soil you need for a desired amount of dry soil:

$$\text{Moist soil wt.} = \text{Dry soil wt.} \times (1 + w_d) \tag{5.12}$$

or

$$\text{Moist soil wt.} = \frac{\text{Dry soil wt.}}{(1 - w_m)}$$
$$= \frac{\text{Dry soil wt.}}{(100\% - w_m\%)} \tag{5.13}$$

5.1.1 Direct Methods of Estimating Soil Moisture and Soil Water Potential

Drying a soil to constant weight at 105°C (sometimes 50–80°C for plant tissue) is the traditional method of arriving at a "dry" sample weight. This temperature is somewhat arbitrary, and clay minerals in particular may contain 10%–15% water (dry basis) at 400°C [42]. As temperature increases, first water in soil pores evaporates, then water adsorbed to mineral surfaces, followed by water between lattice layers and that which forms part of the mineral lattice itself. The exact quantities and patterns of release in a heterogeneous mixture like soil depend on the particular mix of minerals making up a sample. Water adsorbed to organic components (as well as other volatile organic substances) will also evaporate over a range of temperatures. The key point is to specify the temperature used when reporting moisture data.

Instrumental (or indirect) methods of estimation

Physically collecting a moisture sample, besides being labor-intensive and time-consuming, is also very disruptive to the soil. Repeated sampling of a site, particularly at depth, may destroy so much vegetation and soil area as to render the site useless for any other measurements. Instrumental methods, which measure some other property that is related to soil moisture, allow you to monitor soil moisture over time, at various depths, and to automate data collection. As with any instrumental method, good calibration data (comparing instrument readings with gravimetric measurements) is necessary for generating reliable data.

Resistance blocks

Water, especially when it contains dissolved ions, conducts electricity. This fact can be used to derive a relationship between soil water concentration and electrical conductivity. Two wires are embedded in a block of gypsum, nylon, or some other porous material. The block is buried in the soil at the desired depth and is allowed to come to equilibrium with the soil water (which may take days or weeks). A voltage is applied across the free ends of the wires (usually an alternating voltage, to prevent charge polarization of the two electrodes), and the resulting current is measured to indicate resistance (the inverse of conductivity). Precision is low at both the very wet and very dry ends of the scale, and the block may not be truly in equilibrium with the soil water. Gypsum blocks deteriorate with time, but they do supply a steady source of ions which may swamp out variations in soil salinity due to fertilization or irrigation practices.

Neutron thermalization (or neutron probe)

A source of high-energy (or "fast") neutrons (0.1 to 10 Mev) is lowered into an aluminum tube in the soil. The fast neutrons are slowed to "thermal" energy levels by interactions with hydrogen nuclei, found primarily in soil water but also in organic matter. A detector sensitive to thermal neutrons, but not sensitive to fast neutrons, measures the neutron density surrounding the source. This neutron density is correlated to soil water concentration through a calibration curve. One potential drawback could be that other elements found in the soil could contribute to neutron thermalization besides the contribution from water molecules, but for a given soil their effects should remain constant. Another drawback to the method is the spatial resolution, which is at best about 15 cm from the source; i.e., water concentration is averaged throughout a sphere at least 15 cm in radius; variations over shorter distances cannot be detected. Safe handling of the radiation source is imperative.

Time-domain reflectometry (TDR)

When a high-frequency radio pulse is injected into two parallel conductors in a vacuum, it travels along them at the speed of light (3×10^8 ms^{-1}). When it reaches the end it is reflected and travels back to the source. Any matter (air, plastic insulation, or—you guessed it!—wet soil) between or surrounding the conductors slows the pulse velocity and hence increases travel time. The original application for TDR was locating discontinuities (breaks or short circuits) in long power and communications transmission lines, by measuring the round trip travel time of the pulse. Interpretation of TDR data can be tricky, but it is possible to get measurements at specific depths and over a wide range of water concentrations [43].

5.1.2 Procedure to Determine Gravimetric Water Concentrations

Materials Needed for Procedure

1. Sieved soil or ground plant tissue sample
2. Soil cans
3. Soil scoop
4. Balance, displaying to 0.01 g
5. Drying oven (at 105°C for soil or 70–80°C for plant tissue)

Procedure Process

1. Weigh can (with lid) to 0.01 g. Record the can ID number and its weight (Tare weight).

2. Mix sample thoroughly.
3. Place about 10–30 g sample in can. Immediately cover with lid. Note: For samples much less than 10 g, use an analytical balance (0.0001 g precision) and use tongs to handle containers.
4. Weigh and record weight (Tare + Moist Sample).
5. Remove lid and place can (and lid) in oven.
6. After 24 hours, remove cans from oven. Cover with lids and put in desiccator to cool (about 20–30 minutes). Weigh and record weights (Tare + Dry Sample). You may take it out of the oven before 24 hours, but you should check for constant weight (i.e., cool, weigh, put back in oven for an hour, cool, weigh again).

Calculate water concentration, using the appropriate units.

5.2 Soil Bulk Density

Soil bulk density, ρ_b, is defined as the ratio of dry soil mass to bulk soil volume (including pore spaces). The SI unit for density is megagrams per cubic meter ($\mathrm{Mg\ m^{-3}}$), which is numerically equivalent to grams per cubic centimeter.

Just as soil is a combination of soil minerals, organic matter, and air- or water-filled pores, so soil bulk density is a weighted average of the densities of its components:

$$\rho_b = f_a\rho_a + f_p\rho_p + f_o\rho_o + \cdots \tag{5.14}$$

where f is the volume fraction of a component, a stands for air (pores), p stands for soil mineral particles, o stands for organic matter.

Typical densities:

$$\rho_a = 12 \times 10^{-4}\ \mathrm{Mg\ m^{-3}}$$
$$\rho_p = 2.60 - 2.75\ \mathrm{Mg\ m^{-3}} (\text{average value for ``particle density''}$$
$$\text{commonly used is 2.65})$$
$$\rho_o = 0.9 - 1.4\ \mathrm{Mg\ m^{-3}}$$
$$\rho_b = 0.9 - 2.0\ \mathrm{Mg\ m^{-3}}$$

Bulk density is primarily a function of relative pore space and soil mineral particle density as shown in Equation (5.15).

$$\% \text{ pore space} = (1 - \rho_b/\rho_p) \times 100 \tag{5.15}$$

Bulk density is an important soil parameter in its own right, influencing water infiltration and plant root health. It is also an essential piece of information for converting laboratory chemical data, which is commonly expressed on a per unit mass basis (e.g., mg Ca per kg soil), to volume (or area $\times$ depth) units (kg Ca per ha in the 0–30 cm layer).

When calculating bulk density, it is necessary to know whether the mass of soil (the numerator of the fraction) should include the total soil or just the < 2 mm (or "fine earth") fraction; the same is true for soil volume in the denominator. Total mass/total volume would be most useful for construction or engineering applications, where the total mass of material is important. Since all chemical analyses are done on the fine fraction, fine mass/total volume would allow you to convert directly to an area basis for watershed or landscape studies. Fine mass/fine volume, the "true soil bulk density," would tell you what plant roots are experiencing. To determine fine soil volume, Coarse Fraction Volume (CFV) must be subtracted from total volume, which requires assuming a rock particle density (usually 2.65 Mg m^{-3}) as shown in Equation (5.16):

$$\text{CFV(cm}^3) = \text{Coarse fraction mass (g)/Particle density(g cm}^{-3}) \qquad (5.16)$$

To convert soil concentration of an element (for example, 100 mg of exchangeable calcium per kilogram of soil) to soil content (e.g., grams per square meter in the 0–30 cm depth), multiply by the bulk density (with appropriate conversion factors). For example, assuming a bulk density of 1.1 Mg m^{-3},

$$\left(\frac{100 \text{ mg Ca}}{\text{kg soil}}\right) \left(\frac{1.1 \text{ Mg soil}}{\text{m}^3\text{soil}}\right) \left(\frac{1000 \text{ kg soil}}{\text{Mg soil}}\right) \left(\frac{0.3 \text{ m}}{30 \text{ cm depth}}\right) \left(\frac{1 \text{ g Ca}}{1000 \text{ mg Ca}}\right)$$

$$= 33 \text{ g Ca m}^{-2} \text{(in the 0–30 cm layer)}$$

Note: Don't confuse "Mg" as "megagrams" with "Mg" as the symbol for magnesium!

Forest floor is often considered as a whole, rather than as a specific depth. Thus, its bulk density may be expressed in Mass/Area units rather than Mass/Volume (i.e., kg m^{-2}).

5.2.1   Soil Bulk Density Methods

Core method

A cylindrical metal or plastic coring tool of known volume is driven into the soil to a desired depth. The intact core is removed, dried in an oven at 105°C, and weighed.

- Advantages: (1) relatively simple equipment; and (2) undisturbed core.
- Disadvantages: (1) small sampling area of core; (2) stones; and (3) compression of soil inside corer.

A larger diameter core diminishes all these disadvantages, except when rocks are large but closely spaced.

Excavation method

Level soil surface; dig a hole to the desired depth. Line hole with plastic, then fill it with measured volume of water. Excavated soil is dried and weighed.
- Advantages: can be done in stony or gravelly soils.
- Disadvantages: (1) excavated soil is no longer undisturbed; and (2) water gets heavy to lug around.

If done properly, this method will usually give you more accurate numbers than core methods.

Clod method

Coat a clod (a large soil aggregate) with paraffin or other water-repellent substance. Weigh it in air, then in water to determine its volume; or measure the volume of water displaced by the clod in a graduated cylinder.

Radiation methods

Measure radiation transmitted or scattered by soil. This method requires a value of soil water content, and a calibration curve derived from soils with a range of known bulk densities.

Equations

Within a geographical region and for soils of a similar genesis, the close relationship between soil organic matter content and bulk density may allow use of regression equations to calculate ρ_b. For example, Federer [44] derived an equation "$\ln BD = -2.314 - 1.0788 \ln OM - 0.1132 (\ln OM)^2$" for northern New Hampshire till soils. The equation was tested with good results on soils from sites in Maine, southern NH, and Connecticut.

5.3 Soil Texture (Particle Size Analysis or Mechanical Analysis)

Texture, or size distribution of mineral particles (or its associated pore volume), is one of the most important measures of a soil because finely divided soil particles have much greater surface area per unit mass or volume than do coarse particles. Thus, a small amount of fine clay and silt will be much more important in chemical reactions, release of nutrient elements, retention of soil moisture, etc., than a large volume of coarse gravel or sand.

Soil (mineral) particles are broadly segregated into three size classes. (1) **sand**—individual particles visible with the naked eye, (2) **silt**—visible with a light-microscope, and (3) **clay**—some may not be visible with a light-microscope, especially the colloidal size (i.e., < 1 micrometer or 0.001 millimeter). These sand, silt and clay groups are commonly referred to as the **soil separates; soil texture** is defined as the relative proportions of each class and a textural diagram is shown in Figure 3 representing the proportions of sand, silt and clay into the 12 textural classes. However, the specific diameter limits for each class may be different, depending upon the organization making the definition (Fig. 5.1) [45, 46].

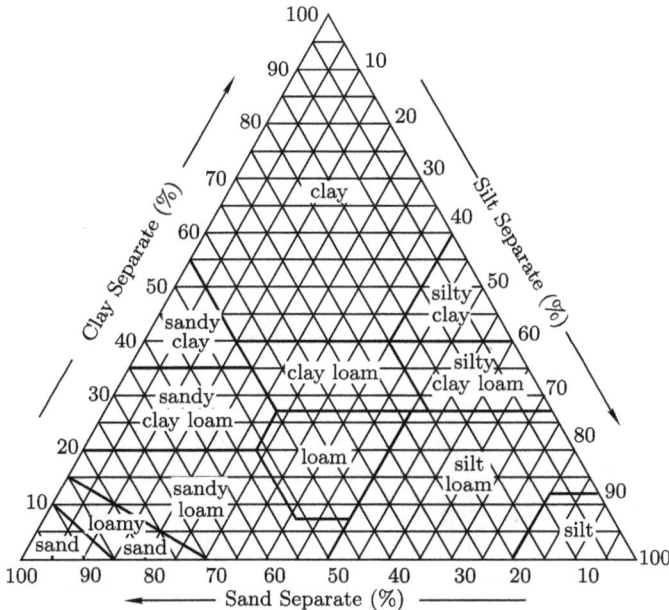

Fig. 5.1 Texture triangle of different proportions of sand, silt, and clay forming 12 texture classes [46].

It is important to recognize two distinct but overlapping uses of the term "clay":

1. The **clay particle size class** includes those particles smaller than 2 micrometers (in the USDA classification).

Relationships Among Particle Size Classes and Different Systems

	FINE EARTH				ROCK FRAGMENTS			150 380 channers flagstones
USDA[1]	Clay	Silt (fi. | co.)	Sand (v.fi. | fi. | med. | co. | v.co.)		Gravel (fi. | med. | co.)		Cobbles Stones Boulders	
U.S.Standard	.002 mm	.02	.05 .1 .25 .5 1		2 mm 5 20 76		250 mm 600 mm	
Sieve No. (opening):			300 140 60 35 18 10	4	(3/4") (3")		(10") (25")	
Inter-national[2]	Clay	Silt	Sand (fi. | co.)		Gravel		Stones	
	.002 mm	.02	.25		2 mm	20 mm		
Unified[3]	Silt or Clay		Sand (fi. | med. | co.)		Gravel (fi. | co.)		Cobbles Boulders	
			.074 .42		2 mm 4.8 19		76 300 mm	
AASHTO[4,5]	Clay	Silt	Sand (fi. | co.)		Gravel or Stones (fi. | med. | co.)		Broken rock (angular), or Boulders (rounded)	
U.S.Standard	.005 mm	.074	.42		2 mm 9.5 25 75 mm			
Sieve No. (opening):		200	40		10 (3/8") (1") (3")			

phi#: 12 10 9 8 7 6 5 4 3 2 1 0 −1 −2 −3 −4 −5 −6 −7 −8 −9 −10 −12

Modified Wentworth[6]	clay — silt — sand — pebbles — cobbles — boulders
U.S.Standard	.002 .004 .008 .016 .031 .062 .125 .25 .5 1 2 mm 8 16 32 64 256 4092 mm
Sieve No. (opening):	230 120 60 35 18 10 5

Fig. 5.2 Comparison of particle size scales according to different organizations [47].

2. The **clay minerals** include particles with mineralogy corresponding to one of the layer aluminosilicate groups (kaolinite, monmorillonite, etc.), and particles of various amorphous or semicrystalline hydrated oxides of iron and aluminum.

Most clay mineral particles are small enough to also fall in the clay particle size class; on the other hand, non-clay minerals (e.g., feldspar or quartz) often dissolve away before reaching such small sizes. The physical and chemical activity of clay in soil is related to both of these features: small particle size is related to **high specific area** (high surface area per unit mass or volume of particles), while mineralogy results in **high surface charge** (high number of charges on the surface) properties.

Soil content of fine particle sizes is very important in determining the fertility and water-supplying capacity as well as tillage characteristics of soils that it is used as one of the primary descriptive characteristics for classifying soil horizons and soil profiles. The U.S. Department of Agriculture soil textural classes are shown on the soil texture triangle (Fig. 5.1).

The major features of particle-size analysis are the destruction or dispersion of soil aggregates into discrete units by **mechanical** or **chemical** means, and then the separation of the soil particles by **sieving** or **sedimentation** methods [45].

Chemical dispersion is accomplished by first removing cementing substances, such as organic matter and iron oxides, and then replacing calcium and magnesium ions (which tend to bind soil particles together into aggregates) with sodium ions (which surround each soil particle with a film of hydrated ions). The calcium and magnesium ions are removed from solution by complexing with oxalate or hexametaphosphate (Calgon) anions [48, 45, 49].

5.3.1   Soil Texture Methods

Commonly used methods to determine soil texture include:

Hand texture method: Qualitative, but with experience many people can easily discern the different textural classes. Kimmins [50] cites some useful criteria for field texturing (Tab. 5.1, Fig. 5.3).

Tab. 5.1 United States Department of Agriculture (USDA) textural classes of soils based on the USDA particle-size classification and potential ranges of percentages of soil separates [50].

Common names of soils (General texture)	Sand (%)	Silt (%)	Clay (%)	Textural class
Sandy soils (Coarse texture)	86–100	0–14	0–10	Sand
	70–86	0–30	0–15	Loamy sand
Loamy soils (Moderately coarse texture)	50–70	0–50	0–20	Sandy loam
Loamy soils (Medium texture)	23–52	28–50	7–27	Loam
	20–50	74–88	0–27	Silty loam
	0–20	88–100	0–12	Silt
Loamy soils (Moderately fine texture)	20–45	15–52	27–40	Clay loam
	45–80	0–28	20–35	Sandy clay loam
	0–20	40–73	27–40	Silty clay loam
Clayey soils (Fine texture)	45–65	0–20	35–55	Sandy clay
	0–20	40–60	40–60	Silty clay
	0–45	0–40	40–100	Clay

Separation by sieving: Mostly used for sand fractionation only (or between 0.05 and 2 mm diameter particles) using American Society for Testing & Materials (ASTM) sieve numbers between 270 or 300 and 10 (openings/inch), respectively. One limitation is that the probability of a particle passing through a sieve in a given time of shaking depends on the nature of the particle, the number of particles of that size, and the properties of the sieve (e.g., particle shape and sieve-opening shape affect probability of passage) [45]. Sieving may be accomplished using either a **wet** (washing type) or **dry** method.

| Moist cast test |

Does the soil ___NO___→ SAND
form a cast?

↓ YES

Does the soil →NO→ Does the soil →NO→ The soil forms a
form a   form a   weak cast
strong cast   weak cast   that does not
that allows   that allows   allow handling
ready handling? careful handing?

↓ YES   ↓ YES   ↓ YES

| Ribbon test |

The soil ←NO─ Does the ←NO─ Does the ←NO─ Does the ←NO─ Does the   Does the →NO→ The soil does
barely   soil form a soil form a soil form a soil form a soil flake   not ribbon.
begins   short and   thin ribbon thin ribbon thin ribbon rather than
to ribbon.   thick   that breaks that holds   longer than ribbon?
              ribbon?   readily and its own   7.5 cm?
                           barely supports weight?
                           own weight?

↓ YES   ↓ YES   ↓ YES   ↓ YES   ↓ YES   ↓ YES   ↓ YES
LOAM   SANDY   CLAY   SANDY LOAM
              CLAY LOAM

| Feel test |

The soil feels ←NO─ Does the   Does the →NO→ The soil has Does the →NO→ The soil   Does the →NO→ The soil feels
moderately   soil feel   soil feel   substantial   soil feel   feels floury soil feel   grainy with a
grainy.   smooth and smooth?   graininess.   very   with slight grainy   considerable
                   floury?   floury?   graininess. with a   amount of
                                                                                     small amount flour material.
                                                                                     of floury
                                                                                     material?

↓ YES   ↓ YES   ↓ YES   ↓ YES   ↓ YES   ↓ YES   ↓ YES   ↓ YES
CLAY LOAM   SILTY   SILTY CLAY   SANDY   SILT   SILT   LOAMY   SILTY
                   CLAY   CLAY   LOAM   SAND   SAND
                   LOAM

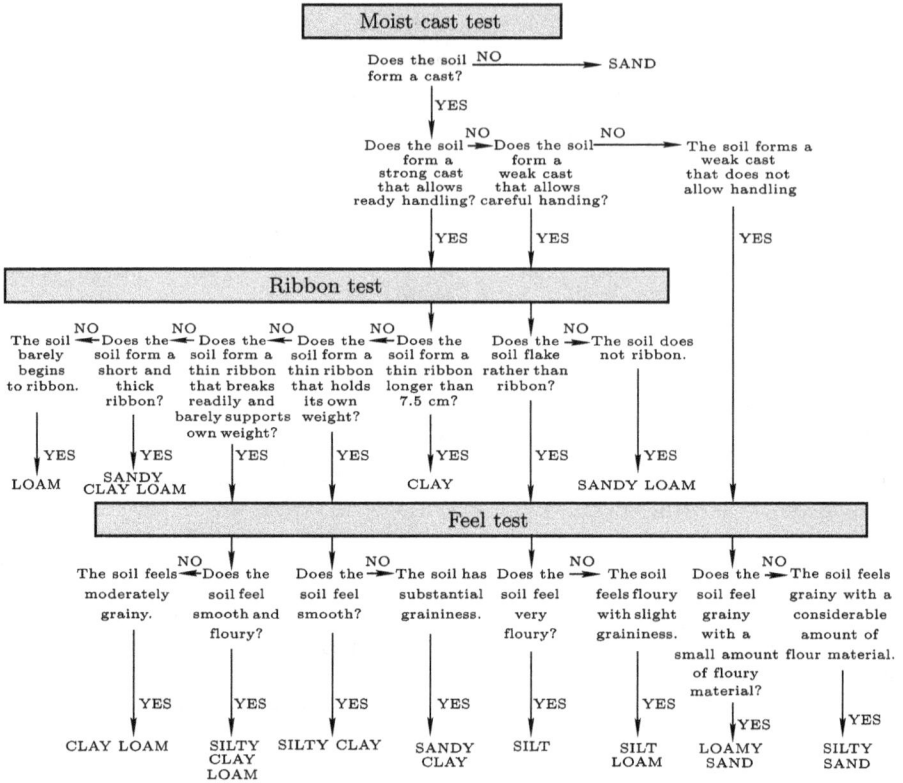

Fig. 5.3 Soil texture classes determined by hand [50].

Separation by sedimentation: This type of analysis depends fundamentally upon **Stokes' Law**. One form of this is shown in Equation (5.17):

$$v = g(\rho_s - \rho_l)\chi^2/(18\eta) \qquad (5.17)$$

where

v = particle velocity of fall;

g = acceleration due to gravity;

ρ_s = particle density;

ρ_l = liquid density;

χ = particle diameter;

η = fluid viscosity.

Basic assumptions used in applying Stokes' Law to sedimenting soil suspensions are:

1. Terminal velocity is attained as soon as settling begins,

2. Settling and resistance are entirely due to the viscosity of the fluid (hydrometer or pipet and the sedimentation-cylinder wall may also influence the settling rate),
3. Particles are smooth and spherical (clay particles especially may be platy),
4. There is no interaction between individual particles in the solution,
5. Ordinarily ρ_s (particle density) is considered to be 2.65 or 2.60 Mg m^{-3} (equivalent to g cm^{-3}), however it may vary between 2.0 to 3.2 Mg m^{-3}, and
6. Temperature of the water should be constant throughout sedimentation.

The **pipet method** is often used as the standard to which other methods are compared. It depends upon the fact that sedimentation eliminates from the depth, h, in a time, t, all particles having settling velocities greater than h/t, while retaining at that depth the original concentration of particles having settling velocities less than h/t. The taking of a small volume element by a pipet at a depth h at time t furnishes a sample from which all particles coarser than χ (particle diameter as determined by Stokes' equation) have been eliminated, and in which all particles finer than that size are present in the same amount as initially. The volume element at depth h has, in effect, been "screened" by sedimentation, so that the ratio of the weight, w, of particles present in that volume at time t, divided by the weight of particles present in it initially, w_0, is equal to $P/100$, where P is the percentage of particles, by weight, smaller than χ. Now, the ratio, w/w_0, can also be written as the concentration ratio, c/c_0, giving $c/c_0 = P/100$. This equation connects the concentration, c, of the pipet sample, in grams per liter, to the parameter P of the particle-size distribution, c_0 being the weight of solids in the entire sample divided by the volume of the suspension [45, 49].

The **Bouyoucos hydrometer method** is somewhat less accurate than the pipet method, but is easier to perform (see Section 5.3.1.1). The theory of the hydrometer method is similar to that of the pipet method except for the manner of determining the concentration of solids in suspension. Letting ρ represent the suspension density, ρ_l the density of liquid, and ρ_s the particle density, all in grams per liter, we have the equation, $\rho = \rho_l + (c/1000)(1 - \rho_l/\rho_s)$. Although the buoyant force on a hydrometer is determined directly by the suspension density (ρ), hydrometer scales can be calibrated in terms of c for particular values of ρ_l and ρ_s. The large size of hydrometer bulb necessary to give adequate sensitivity reduces the depth discrimination of the instrument, but this limitation can be overcome by a simple correction [51].

Depending on the degree of accuracy of separation required, and the particle sizes of interest, the hydrometer method is well adapted for fast determinations of general categories of sizes present and is used in our analyses.

5.3.1.1 Soil Texture Procedure: Bouyoucos Hydrometer Method [1]

Purpose

To measure soil texture by the hydrometer method.

Materials needed for procedure

1. Sieved soil (50 g dry wt. equivalent if fine-textured, 100 g if sandy)
2. Electric mixer and cup
3. Sedimentation cylinder (1,000 mL)
4. Bouyoucos hydrometer
5. Thermometer (e.g., range of $-20°C$ to $110°C$)

Reagents needed for procedure

1. Sodium hexametaphosphate, 1 mol L^{-1}

Procedure process

Note: If soil is not oven dried (OD), take a subsample for water content determination.

1. Place 50–100 g of soil (dry weight equivalent) into a soil dispersing cup. Record the weight to at least 0.1g.
2. Fill cup to within 5 cm of the top with tap water. If local tap water is hard, use distilled water. **Water should be at room temperature, not directly out of tap.**
3. Add 5 mL of 0.167 mol L^{-1} sodium hexametaphosphate.
4. Allow to slake (soak) for 15 minutes (high-clay soils only).
5. Attach cup to mixer; mix 5 minutes for sandy soils, 15 minutes for fine-textured soils.
6. Transfer suspension to sedimentation cylinder; use tap water from squirt bottle to get all of sample from mixing cup.
7. Fill cylinder to 1,000-mL mark with tap water.
8. Carefully mix suspension with plunger. After removing plunger, begin timing. Carefully place hydrometer into suspension; note reading at 40 seconds. This 40-second reading should be repeated several times to improve accuracy. Because the suspension is opaque, read the hydrometer at the top of the meniscus rather than at the bottom.
9. After final 40-second reading, remove hydrometer, carefully lower a thermometer into the suspension and record the temperature (°C). Mixing

---

1 See notes at end for optional pre-treatments to remove organic matter and iron oxide coatings.

raises temperature by 3–5°C, so it is important to record the temperature for both hydrometer readings (40 s and 2 h).

10. Mix suspension again and begin timing for the two-hour reading. Be sure that the cylinder is back from the edge of the counter and in a location where it won't be disturbed.

11. Make up a blank cylinder with water and sodium hexametaphosphate. Record the blank hydrometer reading. If the reading is above 0 (zero) on the hydrometer scale (in other words, if the zero mark is **below** the surface), record the blank correction as a **negative** number. Read at the top of the meniscus as before.

12. Take a hydrometer reading at 2 hours, followed by a temperature reading.

Calculations for % sand, silt and clay (see Eqs. (5.18)–(5.20))

1. Temperature correction factor, T (may be different for each reading):

$$T = (\text{Observed temperature} - 20°C) \times 0.3$$

2. Corrected 40-second reading:

$$40 - s(c) = 40 - s - \text{Blank} + T$$

3. Corrected 2-hour reading:

$$2 - h(c) = 2 - h - \text{Blank} + T$$

4. % Sand $(2 - 0.05 \text{ mm}) = \dfrac{(\text{OD soil wt.}) - (\text{corr. 40 sec reading})}{\text{OD soil wt.}} \times 100\%$

$$(5.18)$$

5. % Clay $(< 0.002 \text{ mm}) = \dfrac{\text{corr. 2 hr reading}}{\text{OD soil wt.}} \times 100\%$ $\qquad (5.19)$

6. % Silt $(0.05 - 0.002 \text{ mm}) = 100\% - (\% \text{ sand} + \% \text{clay})$ $\qquad (5.20)$

Determine your sample's textural class from the textural triangle (see Fig. 5.1).

Optional pre-treatment for soils high in organic matter.

Material needed for procedure

1. 400-mL beaker

Reagents needed for procedure

1. Hydrogen peroxide, 30%
2. Hydrochloric acid, 1 mol L^{-1}

Procedure process

1. Transfer sufficient soil to yield 100 g dry wt., to a 400-mL glass beaker. Add distilled water to give a 1:1 or 1:2 soil-water ratio.
2. If necessary, make the suspension acid to litmus paper with a few drops of 1 mol L^{-1} HCl. Oxidation with H_2O_2 requires an acid medium.
3. Add hydrogen peroxide carefully, a few milliliters at a time, allowing effervescence to subside before adding more, until no more frothing occurs.
4. Place beaker on a hot plate and heat suspension to 65–70°C. When reaction slows, add peroxide slowly as above. Evaporate excess liquid between additions of H_2O_2 to maintain a 1:1 or 1:2 soil-water ratio.
5. Transfer soil to mixing cup, using squirt bottle to get it all out of beaker.

Additional pre-treatment for soils should be considered if the soils to be analyzed are considered to be high in free iron oxides [52].

5.4 Soil Water Potential

Water has many properties that make it essential for life as we know it on Terra. Many of the bulk properties of water (e.g., high specific heat, high surface tension, its ability to dissolve many substances, and the fact that ice floats) are due to the strong attraction of water molecules for each other, through **hydrogen bonding**.

Although a water molecule as a whole is electrically neutral, it is **polarized**: it contains positively and negatively charged regions, in a tetrahedral arrangement. The positive poles arise from the two hydrogen nuclei (protons), whose electrons spend most of their time around the oxygen nucleus. The negative poles result from oxygen's two sets of **lone pair** electrons that are not shared with other atoms. The diagram below (Fig. 5.4) illustrates this arrangement schematically, although a 3-D model best shows the tetrahedral configuration.

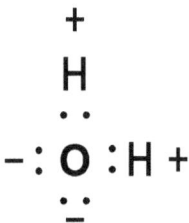

Fig. 5.4 Schematic diagram of water molecule.

Hydrogen bonding strengthens both the attraction of water molecules for each other and the attraction of water molecules for other charged or polarized bodies, including ions in solution, soil particle surfaces, and glass surfaces. These attractions give rise to the concept of the varying **energy state** of water in soil.

Forms of soil water

For most common soils the soil solid phase may occupy from 40% to 70% of the total volume. If the remaining 60% to 30% is filled with water, the soil is said to be saturated. If the pore space is not completely filled with water, but contains air as well as water, then the soil is said to be unsaturated. Traditionally, soil water has been categorized as:

– Gravitational—that water which is in the soil macropores, and which drains from a saturated soil under the influence of gravity alone;
– Capillary—water held in micropores against the force of gravity; and
– Hygroscopic—water closely bound to surfaces of soil particles, even in air-dry or oven-dry soils.

These categories are useful qualitatively as an aid to visualizing the status of soil water. However, a more precise specification of the energy status of water helps to define the actual forces responsible for the movement of water into or out of a particular volume of soil.

Water potential

There are various methods to express the status of water in a soil. One can determine the amount (mass or volume) of water in a soil, or one can determine the energy status of water in a soil. From the point of view of moving water, which always takes place under an energy gradient (e.g., flowing from a high level to a low level), the energy status is more relevant. The same is true if we talk about the **availability of water** for plants. In unsaturated soil, for example, energy has to be applied to remove water from the soil. And, since most plants grow in soil material that is unsaturated, it is therefore valuable to know how much energy must be applied to move water from the soil into the plant.

The energy state of soil water can be described in terms of its motion (**kinetic energy**) and its position in relation to a gravitational, electric, or other force field (**potential energy**). **Kinetic energy**, being related to movement, is equal to one-half the mass of an object times the square of its velocity. In soils the velocity of the water in the pores is rather slow. Thus, in most cases the kinetic energy of water in soils is negligibly small and can, therefore, be safely ignored. (An important exception is in the case of temperature gradients, where differences in temperature result in differences in kinetic energy of water molecules

[53].) In contrast, the **potential energy**, which is related to position in a field (electrical, gravitational, magnetic, etc.), is of great importance. Potential energy of a body at a given point equals the force required to move the body times the distance to that point. It is the potential gradient (the difference in potential energy between two points) that causes water to move from one location to another. Water moves from high potential energy to low potential energy, in the direction of decreasing energy, or "down" an energy gradient.

We are not interested in the absolute amount of potential energy "contained" in the water, but rather in the relative energy in one region compared to that in another. For example, is the potential higher or lower at the soil surface compared to the potential at the bottom of the soil profile?

Several forces act on soil water even at rest, so that **potential energy changes** are important in understanding its behavior. These forces give rise to corresponding **potentials** (defined as potential energy per unit mass or volume of water; when potentials are expressed on a volume basis, units are the same as those of pressure [force per unit area]):

- **Gravity potential**, Ψ_g, due to position above or below a reference elevation
- **Matric (capillary) potential**, Ψ_m, due to attractive forces between water molecules and solid surfaces (**adhesion**), and among water molecules (**cohesion**)
- **Pressure potential**, Ψ_p, due to the weight of water or the pressure of gas above a given point
- **Osmotic (solute) potential**, Ψ_π, due to the presence of dissolved substances.

Other potentials exist, but these four are the most important for our purposes. The total potential of a parcel of water is the sum of the individual potentials.

Potentials must always be expressed relative to a standard **reference state**. For soil water, the reference state is a pool of pure water at a specified elevation and under atmospheric pressure. All the potentials in this reference state are considered to be zero. Then, the following generalizations can be made:

1. Ψ_g can take on any value (positive, negative, or zero), depending on its elevation relative to the reference state. It is often expressed as a **head**, h, in units of length (meters, cm, feet, etc.).
2. Ψ_m is always zero or negative. Soil water above a water table (sometimes called **capillary water**) has a negative matric potential. Ψ_m becomes zero at and below a free water table.
3. Ψ_p can take on any value. It is zero at an infinitesimal distance below a free water surface at standard atmospheric pressure. It increases proportionally

as depth or atmospheric pressure increase. It can be negative in a capillary column (see **surface tension** later).

4. Ψ_π is always zero or negative. Dissolved substances lower the potential energy of water.

The importance of these points can be expressed simply: **water always tends to move in the direction of decreasing total potential** (i.e., it flows downhill or down-gradient, from high to low potential). This fact helps explain many soil phenomena: how water can move upward against gravity in fine pores, or how it can move from a sandy soil into a clay with a higher gravimetric water content (but lower matric potential due to its finer pores).

A corollary to the above statement can also be useful: **If two parcels of water are not moving relative to each other, then their total potentials are equal.**

Advantages to the concept of **soil-water potential** are:

1. It replaces arbitrary definitions, including various "*forms*" of water such as: *gravitational water, capillary water, hygroscopic water*, etc.
2. Water differs from place to place only in potential energy, not in *form*. Values of soil-water potential change gradually and not suddenly.
3. It presents a unified way for describing the status of water in soils, in plants and in the atmosphere.
4. By knowing the difference in potential and its sign between two points in soil, or in the soil and in a plant, we know which way water will move.

Approximate magnitudes of the water potential in the soil-plant-atmosphere system are presented in Table 5.2.

Tab. 5.2 Water potential in the soil-plant-atmosphere system [54].

	Turgid plant (MPa)	Wilting plant (MPa)
Soil	−0.01 to −1.0	−1.0 to −2.0
Leaf	−0.5 to −1.5	−1.5 to −3.0
Atmosphere	−10.1 to −202.7	−10.1 to −202.7

Surface tension

A water molecule in the interior of a parcel of liquid water is attracted to its neighbors in all directions. A molecule at the surface has no neighbors above it: so there is an extra attractive force to be distributed to its neighbors to the sides and below. This extra force at an air-water interface is **surface tension**.

The curvature of an air-water interface at equilibrium is related to the pressure difference across the interface. Pressure difference is zero below a flat surface. When the surface is convex (as in a raindrop), the pressure is greater than atmospheric; below a concave interface (a capillary meniscus) the pressure is negative. This relationship is expressed as in Equation (5.21):

$$\Delta P = 2s/r \qquad (5.21)$$

where

ΔP = the pressure difference across the air/water interface;

s = surface tension (0.073 N/m for water, which is high relative to other liquids); and

r = radius of curvature (negative for concave).

A capillary tube (or soil pore) with a smaller radius will support a greater pressure difference than one with a larger radius. This fact forms the basis for the pressure plate method of establishing specific water potential levels in a soil sample. A porous ceramic plate is saturated with water and subjected to a positive pressure on its upper surface; the lower surface is at atmospheric pressure. Depending on the size of the pores, the surface tension "membrane" at the air-water interface is able to support a pressure difference of up to 15 bars (1,500 kPa). Soil in contact with the plate is initially saturated; over time, water flows out of soil pores in response to the applied pressure gradient. At equilibrium, the only water remaining in the soil is in pores small enough such that the matric potential is equal in magnitude (although opposite in sign) to the pressure potential. Gravimetric water content of the soil is then determined. Now a plot of P vs W_d can be constructed. This plot can then be used as a reference for that specific soil to determine soil water potential at any measured soil water concentration. Since it is much easier to measure W_d in the field than Ψ, such relationships provide a relatively simple means to estimate the energy status of soil water.

Hysteresis

There may not be a strict one-to-one relationship between water content and water potential. In particular, the water content of a soil when it has reached a given potential may depend on whether it reached that point by wetting up from a dry state or drying down from a saturated condition. This condition is an example of **hysteresis**, in which the value of a system variable depends on the direction in which it moved to arrive at that state. When a dry soil is wetted, small pores will fill first with water followed by successively larger pores. Conversely, large pores drain first as a wet soil dries. However, a large pore isolated from the rest of the soil through one or more small interconnecting pores may not be able to drain until the potential has dropped enough to allow the small pores to drain first. Therefore, the drying curve lies above the wetting curve: at a given potential, a drying soil has higher water contents than wetting soil (see Fig. 5.5). Baver *et al.* [48] give other examples of conditions that lead to hysteresis. Note that curves developed from pressure plate measurements (often called moisture release curves) are drying curves, since the soil is initially saturated.

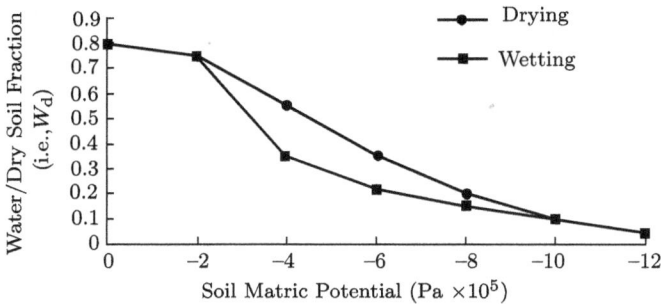

Fig. 5.5 An example of hysteresis in soil water retention curves of wetting and drying curves.

W_d is the gravimetric dry weight ratio (see Eq. (5.1)).

5.4.1 Pressure Plate Apparatus Procedure: Soil Moisture Release Curve

Materials Needed for Procedure

1. Soil samples: one set of replicates for each point to be plotted on the curve. A minimum of 5 points, with three replicates per point, is recommended. Soils may be oven-dried or used moist, or undisturbed cores or sieved, as desired.
2. Rubber retaining rings: typically, about 5 cm I.D. (inner diameter) × 1 cm high.
3. Ceramic plates, rated to cover desired pressure range.
4. Ceramic plate extractors, rated to cover desired pressure range.

Procedure Process

Note: Tap water can usually be used for pressure plate work.
1. Soak plates in water, preferably overnight. All pores must be saturated.

2. Attach outlet tube onto stem (use right-angle adapter on plates going into the 15-bar (1,500 kPa) extractor). Support plate with your hand behind outlet stem when pressing connector in place.

3. Place retaining rings on plate. Make a drawing showing location of samples relative to the outlet stem on the plate.

4. Add soil to retaining rings. Thoroughly mix sample in bag between each spoonful, to minimize particle size separation. Smooth the sample surface even with the top of ring.

5. Prop up the edges of the rubber diaphragm to form a wall around the plate. Carefully add water to the surface of the plate, without disturbing soil samples, so that there is standing water all over the plate. Allow the soil to become saturated by soaking up water from below (if you add water to the top of the soil sample, air bubbles may become trapped in the sample). Cover with waxed paper and let stand until soil is saturated (preferably overnight).

6. Remove excess water from the plates with a syringe (or carefully pour it off). Place first plate in extractor. **The triangular metal spacer must be in the bottom of the 15-bar (1,500 kPa) extractor.** Attach outlet tube to the stem projecting through the chamber wall. Use three plastic spacers to separate plates in the 15-bar extractor; plates in the 5-bar (500 kPa) extractor rest on shelf brackets. Place remaining plates in extractor in the same manner. **Any unused ports must be plugged.**

7. Clean rubber gasket and mating surfaces on lid and extractor body. Close lid and tighten clamping bolts (on 15-bar extractor, tighten two back bolts first, to compress spring that raises up lid).

8. Build up pressure to desired level. **See detailed instructions posted on pressure manifold.**

9. It will take two to three days for the air-soil-water-ceramic plate system to come to equilibrium. To monitor the approach to equilibrium, attach the outlet tube to a burette. Set the initial water level in the burette at the same height as the plates in the extractor, so that no backpressure or extra suction is developed.

10. Record changes in water level over the course of several hours. Note that air bubbles (from air diffusing through the air-water interface) may also change the water level, particularly at high pressures.

11. Weigh empty soil cans, including lid (one for each sample in the extractor).

12. After equilibrium has been established, reduce air pressure in the extractor, following the posted instructions. Open extractor lid, pull outlet tube off of stem, and remove plate.

13. Carefully remove soil and ring from plate and place soil into can. Place lid on can quickly to prevent loss of moisture. Weigh soil + can, put it

in oven at 105°C overnight, then weigh dry soil + can. Calculate water content.

14. Clean plates and retaining rings of any remaining soil, using tap water. Allow plates to dry before storing.

Chapter 6

Soil Chemical Characterization

There are 16–17 macro-nutrients and micro-nutrients needed by plants to grow [16] and the primary source of these elements is the soil. Some exceptions to soils being the source of minerals are: (1) nitrogen which can be derived from both air and soil (+water), however, air has become a significant source mostly from pollution; and (2) oxygen, hydrogen and carbon which are considered non-mineral nutrients because they are not soil derived (and typically are not managed by societies). These essential elements are needed by growing plants and cause human health problems when present in insufficient concentrations in plant tissues that are eaten by humans (Tab. 6.1; Note this is not a complete list of all the human essential elements). Table 6.1 only includes eight of the essential and three of the non-essential elements for plants. All eleven elements listed are essential for humans' to be healthy. Soil nutrients, the element composition of plants, and human health are all interconnected. Humans are only as healthy as the plants ability to take up nutrients from the soil and a plants ability to store these minerals in their tissues. Humans have learned to supplement their diets with those elements that are accumulated at too low concentrations by plants. Low concentrations are found in plants when (1) inadequate levels of these elements exist in soils or (2) because it is a non-essential element for plants so they take up less of it, e.g., iodine is low in some soils such as the mountain regions in Switzerland.

It is worth noting that many of these elements are added to soils as part of fertilizer programs to increase plant growth rates and crop yields. The delivery capacity of soils to provide these nutrients is insufficient to meet the demand of intensely grown crops growing at high growth rates; these growth rates are much higher than natural vegetation growing in the same location. These are essential elements because (1) plants cannot complete its growth cycle and reproduce without it; and (2) no other element can replace the physiological function maintained by each element in a plant.

Some human land-use activities have resulted in either increasing or decreasing the levels of some of these elements in soils, e.g., nitrogen additions due to air pollution or the loss of soil organic matter during long-term intensive agriculture,

Tab. 6.1 Role of some mineral elements in plant growth and functions and mineral links to human health [55, 10].

Mineral	Plant source	Role of element in plants	Element link to human health
Essential elements			
Nitrogen	Soil/Air	Important component of proteins, chlorophyll, nucleic acids	Important component of proteins, amino acids, hormones
Phosphorus	Soil	Component of proteins, coenzymes, nucleic acids; Essential for energy transfer in biochemical pathways (e.g., ADP, ATP)	Essential for energy transfer in biochemical pathways (e.g., ADP, ATP)
Potassium	Soil	Involved with photosynthesis, carbohydrate translocation, protein synthesis, etc.	Helps prevent heart, kidney diseases, diabetes; Deficiencies causes heart disease
Calcium	Soil	Component of plant cell walls; regulates structure and permeability of membranes	Essential for building bones; Deficiencies causes bone deformation
Magnesium	Soil	Enzyme activator; Component of chlorophyll	Deficiencies causes asthma, chronic bronchitis, heart disease, bone deformities
Manganese	Soil	Controls oxidation-reduction systems and photosynthesis	Needed for bone formation; Important for the immune system
Zinc	Soil	Part of enzyme systems regulating metabolic activities	Needed for healthy skin; Maintains immune system; Aids sight/ taste & smell; Deficiencies cause chronic bronchitis
Copper	Soil	Catalyst for respiration; Part of enzymes	Deficiencies cause chronic bronchitis, heart disease & bone deformities
Non-essential plant elements			
Sodium	Soil	Not needed by plants	Deficiencies cause poor nervous system development
Iodine	Soil	Not needed by plants	Deficiencies cause goiters or enlargement of thyroid gland
Selenium	Soil	Not needed by plants	Deficiencies cause chronic bronchitis & heart disease; Small amounts essential for animal & human health

respectively. When these levels increase beyond what is typically delivered by the soil, these concentrations may exist at luxury levels where the soil is incapable of retaining these elements and they leach out into waterways and other lands. Even plants are unable to take up luxury levels of nutrients since plants have a nutrient uptake threshold that is physiologically controlled (Tab. 6.2). This means that increasing fertilizer applications to achieve higher growth rates for plants has thresholds beyond where growth will not increase (Tab. 6.2).

Tab. 6.2 Some nutrient concentration ranges in plants and soils [55, 56].

Element	Thresholds of elements found in plants (mg kg^{-1})	Ranges of elements in some normal soils (mg kg^{-1})
Nitrogen	10,000–50,000	0–1,000
Phosphorus	1,000–5,000	0–40
Calcium	2,000–10,000	0–1,000
Magnesium	1,000–4,000	0–100
Copper	5–20	0–1
Iron	50–250	0–10
Zinc	25–150	0–15

Soils provide a window on nutrient imbalances that alter the ability of the soil and ecosystems [including humans] to be resilient to natural and anthropogenic disturbances. Decreasing nutrient levels, e.g., where nutrients become deficient, or increasing nutrient levels, e.g., where nutrients become toxic to plants and animals, is important to measure to maintain ecosystem health as well as human health. Since not all nutrient elements are taken up by plants, the soil needs to be sampled. During the mid-1970s, the repercussions of acid rain caused the mortality of trees growing in Europe, the Scandinavian countries and the U.S. but the early warning indicators of ecosystem change was detected as changes in soil chemical pools and fluxes [12]. Acidification of the soils and loss of basic cations were shown to cause some tree species to become less resilient following a prolonged drought that ultimately caused tree mortality events. Red Spruce was especially susceptible to acid pollution and drought since it is a species that physiologically is less efficient at taking up soil calcium. Calcium is needed to maintain the integrity of root membranes so they are less permeable to aluminum ions. Red Spruce mortality occurred following the soil chemical changes that made soils more acidic, increased aluminum mobility in the soil and caused the leaching of calcium from the rooting zone. All these changes, especially lower soil Ca levels, resulted in increased aluminum uptake by spruce roots and ultimately aluminum toxicity caused spruce mortality [57, 58].

The texture class of a soil has a significant impact on whether a soil can store and deliver the nutrients needed by plants. For example, sandy soils tend to have lower organic matter levels because decomposition rates are higher, which causes these soils to be nutritionally poor and less able to store water. This is in contrast with soils containing higher clay concentrations, which tend to have higher nutrient and water storage capacity. Clays are also minerals that have charges on their surfaces that attract cations such as ammonium or calcium. This is one of the reasons that analyzing soils for its cation and anion exchange capacity provides important information on how well plants will grow in any soil.

6.1 Soil pH

Perhaps the most important property of soil, as related to plant nutrition, is its **hydrogen ion activity**, or **pH** (the term "reaction" is also used, especially in older literature). Soil reaction is intimately associated with most soil-plant relations. Consequently, the determination of pH has become almost a routine matter in soil studies relating directly or indirectly to plant nutrition. Knowledge of soil acidity is useful in evaluating soils because pH exerts a very strong effect on the solubility and availability of many nutrient elements. It influences nutrient uptake and root growth, and it controls the presence or activity of many microorganisms.

Water dissociates very slightly:

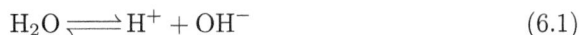

$$H_2O \rightleftharpoons H^+ + OH^- \tag{6.1}$$

The pH scale is based on the ion product of pure water:

$$K_w = [H^+] * [OH^-] = 10^{-14} \text{ at } 23°C \tag{6.2}$$

where K_w is the ion product for water and [ ] indicates the activity of each component in moles per liter of solution. Since $[H^+] = [OH^-]$ in pure water at 23°C, each is equal to $(10^{-14})^{1/2} = 10^{-7}$.

The pH of a solution is defined as the negative log (base 10) of the H^+ ion activity or the log of the reciprocal of $[H^+]$:

$$pH = -\log_{10}[H^+] = \log_{10}(1/[H^+]) \tag{6.3}$$

For example, a hydrogen ion activity of 1/10,000 (or 10^{-4}) mol L^{-1} would equal pH 4. Water with equal numbers of H^+ and OH^- (hydroxyl) is neutral at pH 7 at 23°C. pH values below 7 are increasingly acid with excess H^+ or hydrogen ions. At 100°C the pH of pure water is 6.0 and at 0°C is 7.5 (i.e., temperature affects pH).

Carbon dioxide dissolves in water to form carbonic acid. Otherwise-pure water in equilibrium with CO_2 at its standard atmospheric concentration of 0.033% (330 ppmv) (however it has been suggested to have reached 0.040% in our atmosphere today!) will have a pH of 5.72. But CO_2 concentration may be as high as

10% in poorly aerated soil pores; water in equilibrium with this air would have a pH of 4.45, although other components of soil solution can raise or lower it.

Soil pH

Three soil pH ranges are particularly informative. For example, a pH < 4 indicates the presence of free acids, generally from oxidation of sulfides; a pH < 5.5 suggests the likely occurrence of exchangeable Al; and a pH from 7.8 to 8.2 indicates the presence of $CaCO_3$ [59].

The fundamental property of any acid in general (and therefore of a soil acid) is that of supplying protons, and therefore the H^+ ion activity of a system is fundamentally its proton supplying power. In an analogous fashion, the redox potential (E_h) of a system is its electron supplying power.

Hydrogen ions in solution are in equilibrium with those held on soil particle surfaces (i.e., on exchange sites). The soil pH as actually measured represents the **active** (in solution) hydrogen ion concentration. The total acidity of the soil includes both active and "reserve" (or **exchangeable**) acidity. Thus, two soils with the same pH may have much different amounts of reserve acidity and one may be more difficult to neutralize than another.

Exchangeable aluminum also contributes to soil acidity. When an Al^{3+} ion is displaced from an exchange site into the soil solution, it hydrolyzes, splitting water and releasing a hydrogen ion to solution:

$$Al^{3+} + H_2O \rightleftharpoons AlOH^{2+} + H^+ \qquad (6.4)$$

Lime requirement

The amount of a base (in practice, lime or calcium carbonate) needed to neutralize enough of the exchangeable acidity to raise soil pH to a desired value determines whether a soil is suitable for crop/plant growth.

In most soils it has been noticed that pH levels tend to increase with depth. This is because the upper horizons receive maximum leaching by rainfall, and by dissolved carbonic acid and organic acids which remove metal cations (eg., Ca^{2+}, K^+, Mg^{2+}) and replace them with H^+ ions. Lower horizons are not so strongly leached and, in fact, may accumulate calcium and other materials removed from the upper soil horizons when precipitation rates are low.

6.1.1   Measuring pH

There are many factors that affect soil reaction as measured in the laboratory. The pH of many soils tends to increase as the sample is diluted with water.

Such pH changes may be caused by variables such as carbon dioxide partial pressure, salt concentration, hydrolysis, and solubility of soil constituents. Various soil:water ratios have been proposed for pH determinations. Soil scientists recommend measuring soil pH using very dilute suspensions (1:10 soil:solution ratio) to soil pastes. In general, the pH of most soils increase as the ratio of water to soil increases but this becomes constant at about a soil:water ratio of 1:5.

There is no standard procedure for measuring soil pH. Some of the details that vary from one laboratory to the next are: soil:solution ratio, use of a salt solution (e.g., 0.01 mol L^{-1} $CaCl_2$) rather than water, method of mixing, time of standing before reading, etc. Soil may be weighed, or measured as a volume [60]. Therefore, when reporting soil pH, it is essential to include at least a brief summary of the procedure followed.

The exact placement of the pH electrode in the sample may be important. When placed in the settled sediment of a suspension of soil of appreciable cation exchange capacity (CEC), a lower pH is generally measured compared to the measurement obtained in the supernatant solution (called the suspension effect). However, the sediment pH can be lower than, equal to, or higher than that of the supernatant depending on the soil and existing conditions. For example, if the soil has a net positive charge and more OH^- than H^+ ions are dissociated from the soil, the sediment may have a higher pH than the supernatant [61].

Soil factors in the field that influence soil reaction include degree of base saturation, type of colloid, carbon dioxide partial pressure, oxidation potential, soluble salts, and so on. In addition to these factors the measured pH may vary because of the manner in which the sample is handled in the laboratory before and during the determination. Acquaintance with these variables is necessary for intelligent measurement and interpretation of soil reaction.

A pH can be determined using either **colorimetric** or **electrometric** methods. The choice of method depends upon the accuracy required, the equipment available, or convenience. Many organic dyes are sensitive to pH, the color of the dye changing more or less sharply over a narrow range of H^+ ion activity. These methods tend to be slower, less precise, and obscured from view by soil particles and organic matter. Hence, they are used mostly in the field where pH is to be approximated.

The electrometric method involves a glass electrode that is sensitive to H^+: there is an exchange of ions between solution (H^+) and glass (Na^+) [62] as represented in Figure 6.1. A reference electrode that produces a constant voltage is also required. The electrode pair produces an electromotive force (EMF or voltage) that is measured by a millivoltmeter. The relation between EMF and pH is governed by the **Nernst equation** [63]:

$$E = E_0 + 0.05916\ \text{pH}$$

where
    E = EMF produced by electrode system;
    E_0 = a constant dependent on the electrodes used;
    R = gas constant;
    T = absolute temperature;
    n = number of electrons involved in equilibrium (1 in this case);
    F = Faraday constant.

Note that temperature is a factor in the equation. At 25°C this equation simplifies to

$$E = E_0 + 0.0591 \text{ pH} \tag{6.5}$$

which means a change of 1 pH unit produces a change in EMF of 59.1 mV, at 25°C. This temperature-dependence of pH is important to remember when calibrating a pH meter.

Figure 6.1 describes the parts of the glass electrode used to measure pH:

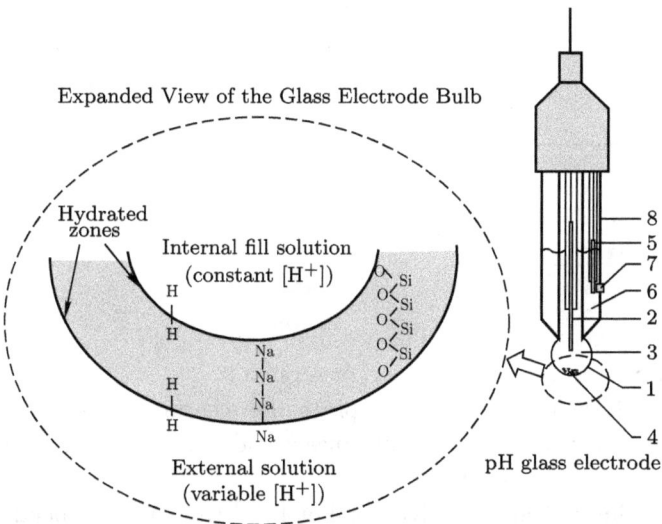

Fig. 6.1 A schematic diagram of a pH glass electrode and expanded view of the electrode bulb (modified from Wikipedia [64]).

where,
1. A sensing part of electrode, a bulb made from specific glass.
2. Sometimes electrodes contain a small amount of AgCl precipitate inside the glass electrode.
3. Internal solution, usually 0.1 mol L^{-1} HCl for pH electrodes or 0.1 mol L^{-1} MeCl for pMe electrodes.

4. Internal electrode, usually silver chloride electrode or calomel electrode.
5. Body of electrode, made from non-conductive glass or plastics.
6. Reference electrode, usually the same type as #4 above.
7. Junction with studied solution, usually made from ceramics or capillary with asbestos or quartz fiber.

Soil reaction classes

The following descriptive terms (Tab. 6.3) are used to characterize specified ranges of soil pH:

Tab. 6.3 Descriptive terms for ranges of soil pH [65].

Soil pH descriptive terms	pH range	Soil pH descriptive terms	pH range
Ultra acid	< 3.5	Neutral	6.6–7.3
Extremely acid	3.5–4.4	Slightly alkaline	7.4–7.8
Very strongly acid	4.5–5.0	Moderately alkaline	7.9–8.4
Strongly acid	5.1–5.5	Strongly alkaline	8.5–9.0
Moderately acid	5.6–6.0	Very strongly alkaline	> 9.0
Slightly acid	6.1–6.5		

6.1.2 The Care of pH Electrodes

– Keep bulb in water/solution as much as possible.
– Don't push bulb into bottom of container, or scratch it.
– Rinse electrode(s) thoroughly between sample/buffer measurements. After rinsing, blot dry—don't wipe, which will cause static charges to build up in the electrode.
– Check level of fill solution (saturated KCl, or 4 mol L^{-1} KCl/AgCl, depending on the type of electrode) in reference electrode; it should be well above the level of the solution you are measuring, to provide sufficient hydrostatic head for a steady flow. Refill if necessary.
– There should be free KCl crystals in the bottom of reference electrode (that's how you know it's a saturated solution). However, be sure crystals are not plugging up ceramic junction.
– Open fill hole when using, to allow free flow of ions during measuring; close fill hole when done, to minimize wasting of fill solution.
– Store combination electrode in 10:1 pH 4 buffer solution:saturated KCl (i.e., dilute the KCl solution with buffer).

– If using separate electrodes: store glass (pH) electrode in pH 4 buffer; store reference electrode in fill solution diluted by 10.

– Because fill solution is flowing out of the reference electrode, you contaminate a sample whenever you place a pH electrode in it, making it unusable for other measurements. In addition to KCl, the fill solution may contain slight amounts of buffer solution that diffuses into it through the porous junction. Commercial buffer solutions often contain mercury as a preservative.

Calibration of pH electrodes

If all of your sample pHs are clustered closely around a single value, you can use a one-point calibration. However, in most cases you should do a two-point calibration, which will allow you to measure a range of pH values. Buffers and samples must be at the same (preferably room) temperature. For most precise calibration, use fresh buffer solution. Rinse electrode thoroughly between each step.

1. Place electrode in pH 7 buffer. Set to 7.00 with **calibrate** control. This control sets the intercept of the pH vs. voltage regression line.

2. Place electrode in pH 4 (or 10) buffer. Set to 4.00 (or 10.00) with **temperature** control. This action sets the slope of the regression line.

3. Repeat steps 1 and 2 until both readings are accurate without changing the controls.

6.2 Electrical Conductivity (EC)

"...the salt of the earth" [66]

The term "soluble salts" refer to the major dissolved inorganic solutes found in soils. Soil salinity is described and characterized in terms of the concentrations (or occasionally the content) of soluble salts (Tab. 6.4). The management and assessment to reclaim saline soils are evaluated based on the measurements of these concentrations (Tab. 6.5). All fertile soils have at least a small amount of soluble salts. Some soil salts are produced in the presence of soil moisture and H_2CO_3 (carbonic acid) equilibrating with exchangeable soil cations. This yields soluble carbonates and bicarbonates of the metallic cations and leaves a corresponding amount of hydrogen ion on the soil exchange complex. Traces to 100 mg kg^{-1} or more of nitrogen can occur as nitrate salts in soils. But the term **soluble salts** is generally considered for soils that have enough salts so that they begin to interfere with normal seed germination, plant growth, or plant uptake of water. When this happens, these soils are termed **saline soils** [67, 68, 69].

Tab. 6.4 Definition of Soil Salinity classes related to Electrical Conductivity (EC) [67–69].

Salinity Class	Electrical Conductivity (EC)(dS/m at 25°C)
Free (of limitations)	0–4
Slight	4–8
Moderate	8–15
Strong	> 15

Tab. 6.5 Soil types related to pH, EC, SAR, and ESP classes [67–69].

Soil	pH	EC (dS/m)	SAR (or ESP)
Normal	6.5–7.2	< 4	< 13(< 15%)
Acid	< 6.5	< 4	< 13(< 15%)
Saline (white alkali)	< 8.5	> 4	< 13(< 15%)
Saline-Sodic	< 8.5	> 4	> 13(> 15%)
Sodic (black alkali)	> 8.5	< 4	> 13(> 15%)

An important consideration to determine is when the actual amount of salt interferes with plant functions, and what is the nature of the particular salt or salts that may be present in a soil. Water-soluble salts in soil consist principally of the four cations—Na^+, K^+, Ca^{2+}, and Mg^{2+}—and five anions—Cl^-, SO_4^{2-}, NO_3^-, HCO_3^- and CO_3^{2-}. Typically, 98% of the soluble salts found in alkali soils consist of these ions.

The negative impact of salinity on plant growth is directly related to the cmol L^{-1} concentration of the soil solution, which is, in turn, related to the total osmotic potential of the solution. The chief effect of the soluble salts is that they decrease the availability of the soil water due to the osmotic potential of the soil solution. A few salts such as borax, and soluble manganese and aluminum have direct toxic effects. Furthermore, plants have different tolerances for some, with some plants tolerating excess salts in soils while others have no tolerance to small amounts of salt in soil solutions. The course of soil development is greatly influenced by the nature (as well as amount) of the salt, and to that extent the nature of the salt is markedly, though indirectly, related to plant growth.

Nutrient concentrations and balance in the soil solution calculated for a field moisture level is utilized to interpret data from analyses of the soil saturation extract. Total salt concentrations and the proportionate amounts of specific ions measured in the soil solution have been correlated with crop responses to determine plant tolerances to these salts as shown in Table 6.6 [70, 71]. Thus, with a single water extract of a soil, we have the capability to evaluate the ionic environment in the soil solution; e.g., the soils total nutrient concentrations and the balances among these nutrients, as well as salinity and pH.

Tab. 6.6 Effects of Electrical Conductivity (EC) on crops [70, 71].

EC (dS/m at 25°C)	EC Effects on some crops
0–2	Salinity effects negligible
2–4	Very sensitive crops affected (citrus, beans)
4–8	Many crops affected
8–16	Only salt tolerant crops yield satisfactorily (wheat, grapes, olives)
> 16	Only very few salt tolerant crops yield satisfactorily (dates, barley, sugar-beets)

To minimize dilution effect, soil soluble salts should be extracted from a soil at moisture contents that approach field levels but also at high enough contents that can be conveniently measured. The saturation extract [69] was chosen for this reason and is obtained by vacuum filtration of a soil paste that has been made up to a saturated condition by adding water while stirring. About 10 mL of the saturation extract is obtained from a soil volume of 75 mL. The electrical resistance of this extract is determined with a conductivity bridge (Fig. 6.2). Electrical conductivity (EC), which is the reciprocal of electrical resistance, varies directly with the soluble salt concentrations. EC readings can be used as an indicator of the soluble salt contents of soils [69].

The most accurate extraction approach to measure soluble salts in soils is to extract the actual soil solution, either through displacement or by means of the pressure membrane apparatus. Extraction made at the soil-saturation moisture content is also a satisfactory way to measure salts because this moisture content is similar to the field moisture content of the soil. Other soil:water extraction ratios widely used are: 1:1, 1:2, and 1:5 extracts as well as the saturated paste extract.

The closer a soil approaches its moisture content to which the roots of plants are exposed, the more directly the soluble salt concentration found may be interpreted in terms of their effects on plants. Since the soil wilting percentage for plants varies over ten-fold, it follows that the soluble salt content of the soil

Electrical Conductivity Cell

Conductivity is proportional to A/L
Conductivity increases with temperature

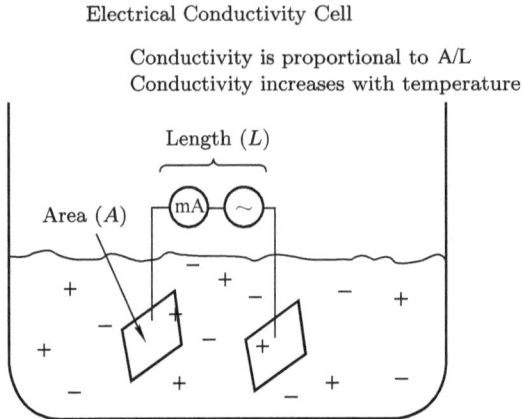

Fig. 6.2 A schematic of an Electrical Conductivity Cell.

solution will become more concentrated in the soil which has the lowest moisture content at a plant's wilting percentage. Thus, in two soils—one a sandy
loam and the other a clayey texture, both will have the same salt content on a
dry soil weight basis. Even though the salt concentration in the soil solution at
the wilting percentage will be approximately 10 times as high in the sandy soil,
because its water content at the wilting point is about 0.1 as great as that of
the clay soil.

For this reason, it is advantageous to extract different soils at different moisture contents, which are directly related to their wilting percentages. The extraction can be made at field moisture contents by means of the pressure membrane apparatus. A more rapid and convenient moisture extract level is the
soil-saturation moisture percentage, with the extract separated by means of a
Buchner funnel. Other, higher moisture contents such as the 1:1, 1:2 or 1:5
soil:water extract ratios can be conducted more quickly, but loses the advantage
of being able to relate the moisture holding properties of the soil, such as the
simple relation of the saturation percentage to the wilting percentage which is
about 4:1 for many soils.

The salt concentrations of soil extracts may be determined by various methods,
including:

1. Gravimetric analysis of total salts through evaporation. Weight of soluble
   salt in dilute solution (%) will possibly give erroneous results due to dissolution of gypsum and calcium carbonate. Weight of soluble salt in saturated
   extract is a more accurate measurement of salt in relation to field conditions but it is more time-consuming than the electrical conductivity (EC)
   determination.

2. Electrical conductivity of saturated paste extract [69]. This method will give reasonable estimates of salt content, but no information of specific ions. It varies with temperature, concentration, and type of salt. For example, EC increases about 2% per °C increase. Ca-, Mg-, and Na-sulfates and $NaHCO_3$ have lower conductivities at the same concentrations than Cl^- salts [69]. A schematic of how an Electrical Conductivity Cell works is presented in Figure 6.2.

3. Total analyses of individual ions of saturated paste extract. This is the most accurate method and results can be converted to mg kg^{-1} (i.e., ppm) or g kg^{-1}, but it is very time-consuming (and expensive).

4. Approximate formulas (see Eq. (6.6)–(6.8) later in this section).

Soluble salts in soils can be determined or estimated from measurements made:
– On aqueous extracts of soil samples,
– On samples of soil solution itself obtained from the soil,
– In soil using buried porous salinity sensors that imbibe and equilibrate with soil water, or
– In soil (or soil pastes) using four-electrode probes and/or electromagnetic induction sensors.

The appropriate method of measuring soil salinity must be selected for the specific condition and purpose. If only a measure of total soluble electrolyte level is needed, salinity sensors, electromagnetic induction and/or four-probe devices are recommended. To monitor soil water salinity as the soil dries between successive applications of irrigation water, the salinity sensor is recommended. When determination of a particular solute is needed, then either the collection or extraction of soil samples or collection of water samples is required. Collection of water samples is more convenient for monitoring needs but is limited to relatively wet soil conditions. Soil sample extracts give relative comparisons only, since the soils are adjusted to unnaturally high water contents during extraction. A combination of the various methods minimizes the need for sample collection and chemical analysis, especially when monitoring salinity changes with time and characterizing large field or project situations. For the latter, the uses of electromagnetic induction and/or four-probe sensors are recommended with supplemental use of the other methods as needed.

Because of marked differences in the equivalent weights, equivalent conductivities, and proportions of major solutes in soil extracts and water samples, the relationships between σ (electrical conductivity) and salt concentration or between σ and Ψ_π (osmotic potential) are only approximate. However, they are still quite useful. These relationships are (for σ in dS/m) are shown in Equations (6.6)–(6.8):

$$\text{Total cation (or anion) concentration, meq/liter} \cong 10 \times \sigma \qquad (6.6)$$

$$\text{Total dissolved solids (TDS), mg/liter} \cong 640 \times \sigma \qquad (6.7)$$

$$\text{Osmotic potential}(\Psi_\pi), \text{bars at} 25°C \cong -0.39 \times \sigma (\text{note negative sign}) \quad (6.8)$$

Conductivity units: The SI unit (International System of Units, or in French, Le Système international d'unités) of conductivity is the siemens per meter (S m^{-1}) (Siemens was an early German electrical experimenter). The old unit, still commonly used, is the millimho per cm or micromho per cm ("mho" is "ohm", the unit of resistance, spelled backward).

$$1 \text{ mmho cm}^{-1} = 1,000 \text{ µmho cm}^{-1} = 0.1 \text{ S m}^{-1} \text{ or } 1 \text{ mmho cm}^{-1} = 1 \text{ dS m}^{-1}$$

6.2.1 Saturated Paste Extract Procedure: Electrical Conductivity

Purpose

To measure EC of a saturated paste extract.

Materials needed for procedure

1. Sieved soil (50–75 g moist wt)
2. 100-mL plastic beaker
3. Plastic "policeman" or glass rod
4. Buchner funnel
5. Whatman No.1 filter paper
6. Vacuum flask
7. Vacuum pump
8. Conductivity meter, with calibration standard (s)

Procedure process

Place 50–75 g of soil into a 100-mL plastic beaker. If you want to determine the moisture content at saturation, weigh the empty beaker plus a glass rod first; add soil and weigh again; then weigh again after saturation is reached.

1. Make a saturated paste, using double distilled (or Nanopure) water.
2. Put filter paper into Buchner funnel. Connect funnel and flask to vacuum pump.
3. Wet filter paper with Double Distilled Water (DDW); turn on pump and open valve.

4. Rinse funnel with DDW.
5. Close vacuum valve; remove funnel, disconnect flask from tube, and rinse flask out with DDW.
6. Re-assemble funnel and flask. Open valve, making sure that filter paper is moist and is drawn down by suction.
7. With the plastic policeman, transfer paste to Buchner funnel. Be sure that no soil gets under filter paper.
8. Filter under suction until no more water is seen to drip from funnel (at least 5 minutes).
9. Close valve, disconnect system, and transfer extract to plastic bottle for EC measurement. (If desired, save extract for later analysis of water-soluble ions.)

Procedure process for quick EC measurement

What follows is a non-standard (but much quicker) method for getting an approximate EC. It will be about one-tenth the value of the saturated paste EC. It may be used to screen a set of samples to find those that warrant the full procedure.

1. Place 2.0 g dry soil in a suitable container,
2. Add 20 mL distilled-deionized water,
3. Stir with a glass stirring rod,
4. Allow soil to settle for a few minutes,
5. Calibrate conductivity meter, and
6. Measure EC in the supernatant liquid.

Conductivity Meter calibration and use

1. If the conductivity cell is dry, it should be soaked in distilled water for ten minutes before use.
2. Carryover from a previous sample, traces of rinse acid from inadequately rinsed glassware, or salt from your skin will contaminate samples. When making a measurement, to minimize contamination of sample or standard: rinse the conductivity cell with distilled water; shake cell to remove as much water as you can; insert cell into a first container of sample as a rinse; then place cell into a second container of sample for final measurement (if you have enough sample for two containers).
3. Gently tap the probe on the bottom of the container to remove air bubbles.
4. Check the calibration in the desired range, using a suitable calibration standard; recalibrate if necessary. Specific instructions for Con 5 meter: after reading stabilizes, press CAL; press the UP or DOWN keys until the displayed value matches the calibration standard value; press ENTER.

5. When finished with the conductivity cell, rinse it thoroughly, then allow it to dry.

6.3 Ion Exchange in Soils

It has been known for centuries that liquid manures lose their color and odor when filtered through soil [72]. Over the years, a number of processes by which soil surfaces (especially of colloidal particles) interact with dissolved substances have been elucidated. Among these processes are those involving surface charge and dissolved ions, or **ion exchange**. Soils may possess both **cation** and **anion exchange capacity.**

6.3.1 Cation Exchange Capacity

Although cation exchange is a universal property of soils, the extent to which different soils are able to exchange cations with solutions varies widely, ranging from a few tenths of a centimole of charge per kilogram of soil ($cmol(+)$ kg^{-1}) to as much as 200 or more $cmol(+)$ kg^{-1}. The highest values are found among peaty soils. The exchange capacity of inorganic soils is rarely more than 75 $cmol(+)$ kg^{-1} and with the vast majority of inorganic soils it is substantially less than 50 $cmol(+)$ kg^{-1}. This exchange capacity is due primarily to organic fractions and clays. However, not all clays contribute equally to this capacity. For example, montmorillonitic clays are generally relatively high in exchange capacity, hydrous micas are intermediate and kaolinitic soils are usually low in exchange capacity. Thus the cation exchange capacity (CEC), the types of exchangeable cations and the chemical and physical properties of these exchangeable cations are dependent to a large extent on the chemical and mineralogical makeup of the soil [73].

Cations (with positive charges) are attracted to negatively charged exchange sites in the soil. The negative charges in soil constituents are derived from a number of sources:

–   Isomorphous substitution within the structures of layer silicate minerals.
–   Broken bonds at mineral edges and external surfaces.
–   Dissociation of acidic functional groups in organic compounds.
–   Preferential adsorption (by chemical reaction) of certain ions on particle surfaces.

The first of these four types of matrix charge is permanent, and is independent of the pH value, the valence of the counter-ion, and the electrolyte level or composition of the bulk solution. The remaining three types of charge vary in magnitude depending on pH, electrolyte concentration, valence of the counter-ion, dielectric constant of the medium, and nature of the anion in the solution phase.

Another source of variable charge in acid soils is that associated with the neutralization of permanent negative charge by strongly adsorbed aluminum-hydroxy polymers that carry a positive charge. As pH rises, these polymers are precipitated as bulk $Al(OH)_3$, thereby freeing the negative sites for participation in normal cation exchange reactions. Negative sites can be similarly neutralized by the adsorption of positively charged mineral particles, such as iron oxides. The positive charges on the oxide/hydroxide surfaces, and their magnitude depends critically on the ionic strength and pH of the solution. Such a charge is substantially neutralized at pH $\geqslant 7$. Another kind of neutralization of permanent charge is that caused by highly selective adsorption associated with the mica silicate minerals, such as biotite, vermiculite, and muscovite, which contain K^+ and NH_4^+ between the contracted platelets. These interlayer cations are not readily exchangeable, although they can be desorbed with certain chemical treatments and through weathering.

Thus, it is obvious that *CEC is a soil property that is dependent upon the conditions under which it is measured.* Different results will be obtained with different methods. Ideally, the method to use is one that measures the soil's capacity to adsorb cations from an aqueous solution of the same pH, ionic strength, dielectric constant, and composition as that encountered in the field, since CEC varies (particularly in certain tropical soils) with these parameters. It is seldom practical to determine the CEC of each soil sample with reagents appropriate to its specific field solution conditions, since the latter information is not easily obtained and each CEC determination would require unique reagents. For this reason CEC determinations are generally based on reference solution conditions that must be standardized to obtain data that can be applied and interpreted universally. **The method used should always be reported with the data** (for example, the saturating cation used and the pH of the saturating cation solution).

Many methods for determining CEC are provided by using different combinations of soil pretreatment, saturation, washing, and extraction procedures, and different saturation and replacing of cations, washing solvents, and pH control. Most methods used may be categorized as one of five:

1. The exchangeable cations can be displaced with a saturating salt solution and the CEC taken as equivalent to the sum of exchangeable cations present in the reacted "leachate" (**summation method**; the resultant sum is also called **effective cation exchange capacity or ECEC**).

2. After the CEC has been saturated with an index cation, the adsorbed cation and the small amount of solution entrained by the soil (solution in the soil

pores) after centrifuging can be displaced directly by another salt solution without further treatment of the soil. The saturating cation and anion are then determined in the resulting extract, and their difference (since the anion is only present in the entrained solution) is taken as equal to the CEC of the soil (**direct displacement**).

3. When the exchange sites have been saturated with an index cation, the soil can be washed free of excess saturating salt, and the index cation adsorbed by the soil can then be displaced and determined (**displacement after washing method**).

4. When the exchange sites have been saturated with a cation such as barium, this adsorbed cation can be displaced by a solution whose anion forms an insoluble precipitate with it (e.g., $MgSO_4$). This step removes the initial adsorbed cation without having to wash it out. CEC is then determined by the decrease in the concentration of magnesium remaining in solution, after it has displaced barium on the exchange complex (**compulsive exchange method**) [74].

5. Following saturation of the soil CEC with an index cation, the saturating solution can be diluted and labeled with a radioactive isotope of the saturating cation. The concentration of the index cation in the solution is then determined, and the distribution of the isotope (and hence of the total cation) between the two phases is given by measuring the radiation in the solution and soil plus solution (**radioactive tracer method**).

Variations in results are not surprising in view of the many possible complicating interactions between saturating, washing, and extracting solutions and soil constituents during the analysis and the fact that CEC is not an independent, single-valued soil property. The complications arising from the dissolution of $CaCO_3$ and gypsum and the presence of salt in the soil during CEC determination are particularly troublesome for arid-land soils. Determination of CEC of acid soils, on the other hand, is complicated because of their variable charge character and relatively high content of the more difficultly exchangeable aluminum-hydroxy "cations". For these reasons, different methods of CEC determination are recommended for arid and acid soils. For arid land soils, a modification of the method of Polemio [75] is recommended. For acid soils, the methods of Gillman [76] and Sumner and Miller [77] are recommended.

Ammonium acetate (buffered at pH 7) and sodium acetate (pH 8.2) have been employed widely for determining soil CEC. Significant errors result when CEC is determined with these methods on soils that contain calcium carbonate, gypsum, zeolites, feldspathoids, or vermiculite minerals. These methods are described by Chapman [78].

Either two or three steps are commonly used in the conventional methods of determining the CEC of soils, but potential **errors** exist in each step. The three steps are (1) saturation of cation exchange sites with a specific cation, (2) removal of excess saturating solution (this step is eliminated in two-step methods), and (3) replacement of saturating cation. Possible sources of error in these steps include the following:

– In the saturation step, the exchange sites may not be completely saturated with the saturating cation because other cations in the saturating solution compete for adsorption sites or because the saturating cation's replacing power is insufficient to replace the more strongly adsorbed cations (e.g., exchangeable Al and its hydroxy forms). Other cations may be present in the saturating solution because of the dissolution of $CaCO_3$, gypsum, and silicate minerals during saturation. Such dissolution may be appreciable in certain commonly used saturating solutions [75, 79]. Exchangeable Al and its hydroxy forms are not readily exchanged with monovalent cation saturating solutions. This error would result in an underestimate of CEC.

– In the washing step, there are four potential sources of error. The adsorbed cation may be removed by hydrolysis and replaced by the H^+ ion. It may be replaced by cations (especially Ca^{2+}) brought into solution in the washing solvent from the dissolution of $CaCO_3$, gypsum, and silicates. Cation exchangers (especially fine clay particles and organic matter) may be lost during decantation because they tend to disperse as the excess electrolyte is removed during washing. Some of the original saturating solution may be retained in the sediment and subsequently extracted as an exchangeable cation if the washing is incomplete or if salt is retained. All but the last of these sources of error cause an underestimate of CEC.

– In the replacement step, there are two potential sources of error. First, the adsorbed cation may be trapped between interlayers by contraction of expandable 2:1 layer silicates (especially vermiculites and weathered micas) if the replacing solution contains NH_4^+ or K^+. This entraps the saturating cation and prevents its replacement during extraction. This is a common problem with many arid land soils [80] and results in an underestimate of CEC. Second, nonexchangeable cations may be extracted from zeolite, feldspathoid, feldspar, and mafic minerals by the replacing solution. This error is also common with arid land soils [79], especially if Ca^{2+} or Mg^{2+} is the saturating cation, if NH_4OAc is the replacing solution, and if the soils are calcareous, gypsiferous, and relatively unweathered. This error results in high CEC values.

The relative ability of one cation to displace another cation from an exchange site varies, primarily with the ion's **valence** and with its **hydrated radius**

[81]. Ions with a higher valence are retained more strongly than those with a lower valence; so $Al^{3+} > Ca^{2+} > K^+$. Within a given valence series (for instance, going down a group in the periodic table), smaller ions have a greater density of charge (the same charge is spread over a smaller volume). Smaller ions in solution thus attract water molecules more strongly and in fact have a larger **hydrated radius** than do larger ions of the same valence. Since the hydrated radius determines how close an ion can get to a charged particle surface (and therefore how tightly it is attracted), an ion with a smaller hydrated radius (lower in the periodic table) is more strongly retained: $Rb^+ > K^+ > Na^+$. The combination of these two trends produces the **lyotropic** (or Hofmeister) **series**, given in order of decreasing attraction to exchange sites:

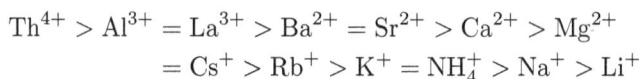

$$Th^{4+} > Al^{3+} = La^{3+} > Ba^{2+} = Sr^{2+} > Ca^{2+} > Mg^{2+}$$
$$= Cs^+ > Rb^+ > K^+ = NH_4^+ > Na^+ > Li^+$$

The large hydrated radius of sodium essentially accounts for its role in dispersing clay particles.

6.3.2   Exchangeable Cations

Exchangeable cations are those that can be readily exchanged by the cation of an added salt solution. In many cases, the definition of an exchangeable cation is entirely straightforward, that is, any added cation will exchange with the adsorbed soil cation. In other cases, the definition is less precise. Examples of these quasi-exchangeable cations are K^+ in micaceous soils, Al^{3+} in acid soils, and H^+ in practically all soils.

The ratio of exchangeable **base cations** (Ca^{2+}, Mg^{2+}, K^+, and Na^+) to total CEC is called **base saturation.** For consistency, you may want to use the effective CEC (ECEC, the sum of exchangeable cations). The formula may then be written as

$$\text{Base saturation} = \left(\frac{\sum \text{Exch. bases}}{\sum \text{Exch. bases} + \text{Exch. aluminum}} \right) \times 100\% \qquad (6.9)$$

In most agricultural soils, Ca^{2+} occurs in larger quantities than all other bases combined. In some soils formed from serpentine, Mg^{2+} may be more abundant than Ca^{2+}. This tendency is also found in acid subsoils on older landscapes. Calcium is low because there is no continuing source of it in the soil minerals, but Mg^{2+} is high because it is dissolved from clays under acid conditions. Potassium is usually the third most abundant base and varies because of both soil parent material and agricultural treatments. Sodium in most soils is very low and in many humid region soils approaches only traces. In natric soils, however, Na^+

may be present at a concentration of several cmol kg^{-1} (or meq 100 g^{-1}). Even under these conditions, it rarely is present in as high a concentration as Ca^{2+}. The monovalent base cations (K^+ and Na^+) are much more readily leached from soils than are the divalent bases (Ca^{2+} and Mg^{2+}).

Other common exchangeable cations include Al^{3+} and H^+, the primary ions of **exchangeable acidity**. Other cations that may be analyzed as to their abundance on the soil exchange sites include Mn^{2+}, Fe^{3+}, Cu^{2+}, Zn^{2+}, Co^{2+}, Pb^{2+}, etc. These cations are often involved in hazardous waste disposal issues. Their presence and quantity are especially needed to model biological processes in or above the soil. These processes may be either positively or negatively influenced by metal concentrations on exchange sites. However, many of these multivalent metal ions interact strongly with complex organic molecules found in soil solution, and thus may not be present in an exchangeable ionic form. They may also form insoluble oxides or hydroxides, particularly at higher soil pH.

6.3.3   Extraction Procedures for Exchangeable Cations and Cation Exchange Capacity

A number of salt solutions may be used for displacing exchangeable ions from soil [77, 82]. Commonly used are 1 mol L^{-1} ammonium chloride, 0.25 or 0.5 mol L^{-1} barium chloride, and 1M ammonium acetate (buffered to pH 7.0 or 8.1). The choice of solution depends on a number of considerations:

- The salt ions should not interfere with subsequent analysis of target ions in the extract.
- The extracting solution should not react with soil components. For example, barium salts are not appropriate for arid land soils because Ba precipitates with carbonates and sulfates.
- A solution buffered at a neutral or alkaline pH will help minimize dissolution of carbonates in alkaline soils, but unbuffered solutions work best for acid soils, where pH-dependent charges may be present.
- Solution concentration should be high enough to completely displace target ions from the exchange complex. One molar solutions are commonly used; because of its position in the lyotropic series, barium can be used at lower concentrations.
- Cost may be a factor, particularly for disposal of hazardous wastes such as barium.

Units: The traditional unit for exchange capacity, as well as for individual exchangeable ions, is **meq 100 g^{-1} soil**, which typically has a value between

0.1 and 100. The SI unit is **cmol(+) kg^{-1} soil** (or cmol(−) kg^{-1}, for anions), **which is numerically equal to the older non-SI unit.**

Conversions: When converting an analytical concentration (such as mg Ca per L of extract) to equivalent units (cmol(+) kg^{-1} soil), you multiply by the valence and divide by the atomic weight. For example (as in Eq. (6.10)), if a 2.00-gram sample extracted with 20 mL of solution has a calcium concentration of 25 mg Ca L^{-1}:

$$\left(\frac{25 \text{ mg Ca}}{\text{L extract}}\right)\left(\frac{0.02 \text{ L extract}}{0.002 \text{ kg soil}}\right)\left(\frac{1 \text{ mmol Ca}}{40.08 \text{ mg Ca}}\right)$$

$$\left(\frac{2 \text{ meq Ca}^{2+}}{\text{mmol Ca}}\right)\left(\frac{1 \text{ cmol(+)}}{10 \text{ meq Ca}^{2+}}\right) = 1.25 \text{ cmol(+)/kg soil} \qquad (6.10)$$

Purpose

To measure exchangeable cations and cation exchange capacity by the compulsive exchange method [77].

Materials needed for procedure

1. Sieved soil (1.00 g dry wt. equivalent for fine-textured/high Organic Matter (OM) soils; 2.00 g dry wt. equivalent for coarse/low OM soils. Use 1.50 or 3.00 g moist sieved soil if you don't know the actual moisture content, and take a subsample for moisture determination.)
2. 50 mL plastic centrifuge tubes.
3. Table shaker.
4. Centrifuge.
5. Barium chloride/ammonium chloride extracting solution (0.2 mol L^{-1}): 48.9 g BaCl$_2$·2H$_2$O+10.7 g NH$_4$Cl per L.
6. 0.05 mol L^{-1} barium chloride solution: 12.2 g BaCl$_2$·2H$_2$O per L.
7. Equilibrating solution, 0.002 mol L^{-1} barium chloride: 0.4889 g BaCl$_2$·2H$_2$O per L (or 40 mL of 0.05 mol L^{-1} solution per L).
8. Magnesium sulfate heptahydrate solution (0.005 mol L^{-1}): 1.2324 g MgSO$_4$ ·2H$_2$O per L.

Procedure process

Note: (1) At least one blank (no soil) **must** be run with each set of samples. This blank value is required for CEC calculations, as well as for normal Quality Assurance/Quality Control (QA/QC) purposes (Appendix B). (2) **Barium salts are hazardous materials** and should not be poured down the sink. Place waste barium solutions in a bottle for proper disposal. (3) If you only wish to extract exchangeable cations, without measuring CEC, then perform Steps 1 through 5 only.

1. Record weight of empty centrifuge tube (to 0.01 g), including cap.
2. Place soil in tube (2.00 g dry for sub-surface, 1.00 g dry for surface/high OM). Add 20.0 mL of 0.2 mol L^{-1} $BaCl_2$/NH_4Cl extracting solution.
3. Cap tube. Shake the tubes for 1 hour.
4. Centrifuge tubes at 2,000 $\times g$ for 5 minutes (remember to balance centrifuge rotor).
5. Decant supernatant liquid (may be saved for exchangeable cation/acidity analysis).
6. Add 20.0 mL 0.05 mol L^{-1} $BaCl_2$. Disperse solids on bottom with a mixer (e.g., with a vortex mixer). Shake for 10 minutes.
7. Centrifuge for 5 minutes. Decant and discard supernatant.
8. Add 20.0 mL 0.002 mol L^{-1} $BaCl_2$. Disperse solids on bottom with vortex mixer. Shake for 10 minutes.
9. Centrifuge for 5 minutes. Decant and discard supernatant.
10. Repeat steps 8 and 9 two times.
11. After decanting 0.002 mol L^{-1} $BaCl_2$ for the third time, weigh tube plus soil plus entrained solution.
12. Add 20.0 mL 0.005 mol L^{-1} $MgSO_4$. (The amount of this solution can be varied somewhat to accommodate lower or higher CECs. 10 mL corresponds to 2 g soil with a CEC of 5 cmol(+) kg^{-1}.) Disperse solids. Shake 30 minutes. **Note:** For critical determinations, the solution ionic strength can be adjusted at this point by matching its electrical conductivity to that of a reference solution. See Gillman [76] and Sumner and Miller [77] for a description.
13. Let tubes stand overnight, with occasional hand shaking. This long period of equilibration is necessary to allow Mg^{2+} ions to completely displace Ba^{2+} on the exchange complex. Displaced barium ions precipitate as $BaSO_4$ and are thus removed from further exchange reactions.
14. Centrifuge 10 minutes. Decant solution into a labeled bottle. Analyze on Inductively Coupled Plasma (ICP) spectroscopy for [Mg], and [Ba] as a check on the method. Note that barium sulfate is very slightly soluble; the maximum concentration of Ba over precipitated $BaSO_4$ is about 1.5 mg L^{-1}—high Ba indicates insufficient Mg for complete displacement. If high Ba, decrease the amount of soil used or increase $MgSO_4$ levels.

Calculations

The total CEC is equivalent to the mass (in millequivalents) of Mg^{2+} removed from solution, divided by the dry soil weight. If moist soil was used, be sure to calculate the equivalent dry weight.

1. Calculate **entrained** solution volume:

$$\text{Entrained} = (\text{wt. of tube} + \text{soil} + \text{entrained})$$
$$-(\text{wt. of empty tube} + \text{oven-dry soil}) \quad (6.11)$$

This number represents the **mass** of entrained solution; assuming a specific gravity of 1 (close enough for a very dilute solution), it is also the **volume**, in mL.

2. Convert **concentration** in the extract (mg Mg L^{-1}) to total **mass** (total meq Mg^{2+}), **for both the blank and sample** (remember, Mass = Concentration $*$ Volume; this step also incorporates unit conversions for mg to meq and mL to L). The formula below (Eq. (6.12)) assumes that you added 20 mL of 0.005 mol L^{-1} $MgSO_4$; substitute the appropriate value if necessary:

$$\text{meq } Mg^{2+} = \frac{??\text{mg Mg}}{L} \times \frac{1 \text{ mmol Mg}}{24.305 \text{ mg Mg}} \times \frac{2 \text{ meq } Mg^{2+}}{1 \text{ mmol Mg}}$$
$$\times (20 \text{ mL} + \text{entrained}) \times \frac{1 \text{ L}}{1,000 \text{ mL}} \quad (6.12)$$

3. The **difference** between meq Mg^{2+} in the blank and in the sample represents the mass of magnesium ions adsorbed onto the soil exchange complex. (No magnesium ions were removed from the blank solution, so meq Mg^{2+} measured in the blank represents what was originally added to the sample tube, before some of it was removed by soil CEC.) Divide by (dry) soil mass, convert meq to cmol(+) and g to kg, and you have CEC as shown in Equation (6.13):

$$CEC(\text{cmol}(+)/\text{kg soil})$$
$$= \frac{(\text{meq } Mg^{2+}_{\text{blank}} - \text{meq } Mg^{2+}_{\text{sample}})}{2 \text{ g moist soil}} \times \frac{(1 + w_d)\text{g moist soil}}{\text{g dry soil}}$$
$$\times \frac{1 \text{ cmol}(+)}{10 \text{ meq } Mg^{2+}} \times \frac{1,000 \text{ g soil}}{\text{kg soil}} \quad (6.13)$$

A note on units

The traditional unit for exchange capacity, as well as for individual exchangeable ions, is **?? meq 100 g^{-1} soil**, which typically has a value between 0.1 and 100. The SI unit is **cmol(+) kg^{-1} soil** (or ?? cmol(−) kg^{-1}, for anions), **which is numerically equal to the older unit.**

When converting an analytical concentration (such as mg Ca per L of extract) to equivalent units (cmol(+) kg^{-1} soil), you multiply by the valence and divide by the atomic weight. For example, a 2.00-gram sample extracted with 20 mL of solution contains 25 mg Ca L^{-1}:

$$\left(\frac{25 \text{ mg Ca}}{\text{L extract}}\right)\left(\frac{0.02 \text{ L extract}}{0.002 \text{ kg soil}}\right)\left(\frac{1 \text{ mmol Ca}}{40.08 \text{ mg Ca}}\right)$$

$$\left(\frac{2 \text{ meq Ca}^{2+}}{\text{mmol Ca}}\right)\left(\frac{1 \text{ cmol}(+)}{10 \text{ meq Ca}^{2+}}\right) = 1.25 \text{ cmol}(+)/\text{kg soil}$$

6.4   Exchangeable Soil Acidity

Exchangeable acidity (EA) in soil is a fairly arbitrary quantity derived from several sources [83]: Hydrogen ions can come from the hydrolysis of exchangeable, trivalent Al:

[1] $Al^{3+}+3H_2O \longrightarrow Al(OH)_3+3H^+$

They may also come from hydrolysis of partially hydrolyzed and nonexchangeable Al.

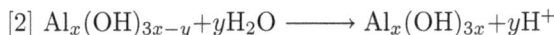

[2] $Al_x(OH)_{3x-y}+yH_2O \longrightarrow Al_x(OH)_{3x}+yH^+$

A third type is from weakly acidic groups, mostly on organic matter:

$$[3] \quad R-\overset{\overset{\displaystyle O}{\|}}{C}-OH \longrightarrow R-\overset{\overset{\displaystyle O}{\|}}{C}-O^- + H^+$$

and a fourth is exchangeable H$^+$:

[4] $XH \longrightarrow X^-+H^+$

where X$^-$ represents an exchange site on a soil colloid.

Reactions [2] and [3] are the most important in soils with pH > 5.5, with reaction [1] being important below pH 5.5 and reaction [4] important below pH 4. In other words, free hydrogen ions exist on soil exchange sites only in very acid soils. In most soils, aluminum and organic matter constitute the exchangeable or **reserve acidity**, which is in equilibrium with H$^+$ and Al^{3+} in solution (the **active acidity**). It is the active soil acidity that plant roots respond to (and that pH electrodes sense). However, the exchangeable acidity is much larger. In order to raise soil pH (by adding limestone, for example), the reservoir of exchangeable acidity must be neutralized, not just the relatively small quantity of hydrogen ions in solution.

Exchangeable acidity can be measured using either buffered or unbuffered extractants [83]. The buffer may be more appropriate where soil pH adjustment is required, as in liming of an agricultural soil, while the unbuffered salt may be better in an acid forest soil.

Exchangeable acidity may also be estimated by determining Total CEC, summing the base concentrations (Ca^{2+}, Mg^{2+}, K^+, and Na^+) in the CEC leachate, and taking the difference as EA. One problem with this method is that all five measurements have individual errors, which are combined by summing and subtracting.

6.4.1 Extraction Procedures for Exchangeable Soil Acidity

Concentrated unbuffered salt solutions (such as 1 mol L^{-1} KCl or 0.5 mol L^{-1} $BaCl_2$) are most suited for estimating "truly" exchangeable aluminum, that which most closely approximates the fraction related to plant response [84]. Acid salt solutions may begin to displace Al from "nonexchangeable" sources, such as structural hydroxy-aluminum polymers.

6.4.1.1 Exchangeable Acidity (Barium Chloride—Triethanolamine Method)

Purpose

To measure exchangeable acidity by a buffered extracting solution [83].

Materials needed for procedure

1. Sieved soil
2. Plastic beakers (50- or 100- mL)
3. Glass stirring rod
4. Erlenmeyer flasks (250- or 500-mL)
5. Buchner funnel
6. Whatman No.42 filter paper
7. Vacuum flask
8. Vacuum pump
9. Titration setup
10. Soda lime trap

Reagents needed for procedure

1. Barium chloride dihydrate 0.25 mol L^{-1} + triethanolamine (TEA) 0.2 mol L^{-1} buffer solution: dissolve 61.07 g $BaCl_2 \cdot 2H_2O$ L^{-1} + 29.8 g TEA L^{-1} in deionized water; adjust to pH 8.2 with hydrochloric acid (HCl). Protect

from CO_2 contamination by attaching a tube containing soda lime to the air intake.

2. Barium chloride replacing solution: dissolve 61.07 g $BaCl_2 \cdot 2H_2O$ L^{-1} of deionized water; add 0.4 mL of buffer (buffer solution in #1 above) L^{-1}. Protect from CO_2 as above.
3. Hydrochloric acid, approximately 0.2 mol L^{-1}, standardized for titration.
4. Bromocresol green, 0.1% aqueous solution.
5. Mixed indicator: dissolve 1.250 g methyl red and 0.825 g methylene blue in 1 L 90% ethanol.

Procedure process

1. Weigh 10.0 g soil (dry weight equivalent; or use 5.0 g dry weight equivalent for soils high in acidity) into a plastic beaker. Add 25.0 mL buffer solution. Stir with glass rod to mix.
2. Let stand for 1 hour.
3. Transfer to Buchner funnel fitted with No. 42 filter paper. Filter into the Erlenmeyer flask.
4. Add 75 additional mL of buffer solution, in 3 25-mL increments (you may use these to rinse all of the soil particles from the beaker).
5. Run 100 mL of replacing solution through the filter.
6. Add 2 drops of bromocresol green and 10 drops of the mixed indicator to the flask.
7. Titrate with standard HCl to an endpoint in the range from green to purple (titrate all samples to the same endpoint).
8. Prepare a blank solution from 100 mL of buffer and 100 mL of replacing solution; titrate as above, to same endpoint.

6.4.1.2   Exchangeable Acidity (Potassium Chloride Method)

Purpose

To measure exchangeable acidity by an unbuffered extracting solution [83].

Materials needed for procedure

1. Sieved soil
2. Plastic beakers (50- or 100-mL)
3. Glass stirring rod
4. Erlenmeyer flasks (250- or 500-mL)
5. Buchner funnel
6. Whatman No.42 filter paper

7. Vacuum flask
8. Vacuum pump
9. Titration setup

Reagents needed for procedure

1. Potassium chloride replacing solution, 1M: Dissolve 74.56 g KCl L^{-1} of deionized water.
2. Potassium fluoride, 1M (aluminum complexing solution): dissolve 58.1 g KF L^{-1} of deionized water; titrate to a phenolphthalein endpoint with sodium hydroxide (NaOH).
3. Hydrochloric acid (HCl), approximately 0.1M, standardized for titration.
4. Sodium hydroxide, approximately 0.1M, standardized for titration.
5. Phenolphthalein solution: dissolve 1 g phenolphthalein in 100 mL ethanol.

Procedure process

1. Weigh 10.0 g soil (dry weight equivalent; use 5.0 g for soils high in acidity) into a plastic beaker. Add 25.0 mL KCl solution. Stir with glass rod to mix.
2. Let stand for 30 minutes.
3. Transfer to Buchner funnel fitted with No. 42 filter paper. Filter into the Erlenmeyer flask.
4. Add 125 additional mL of KCl solution, in 5 25-mL increments (you may use these to rinse all of the soil particles from the beaker).
5. Add 4 or 5 drops of phenolphthalein indicator to the flask.
6. Titrate with standard NaOH to the first permanent pink endpoint (if the color is deep pink, it is too far). This gives the KCl acidity.
7. To estimate separate amounts of Al^{3+} and H$^+$, record the amount of standard NaOH added, add 10 mL of 1M KF, and titrate with 0.1M HCl until pink color disappears.
8. Wait 30 minutes.
9. Add additional HCl until pink color disappears.
10. Prepare a blank solution from 150 mL of KCl solution; titrate with NaOH as above, to same endpoint.

Calculations

KCl acidity (cmol(+)/kg soil) = [(mL NaOH for sample − mL NaOH for blank) × ??mol L^{-1} NaOH × 100] / g sample
KCl exchangeable Al (cmol(+)/kg soil) = (mL HCl × ??mol L^{-1} HCl × 100) / g sample
KCl exchangeable H (cmol(+)/kg soil) = KCL acidity − KCL exchangeable Al

6.5   Extractable Inorganic Soil Nitrogen

Nitrogen analysis is performed to learn the comparative nitrogen status of soils or soil horizons as some indication of its availability for plants (crops or trees). Most commonly total nitrogen content is determined on a percentage basis and may range from a few hundredths of a percent in infertile mineral soils to more than 1% in some organic horizons or soils. Total amounts in forest soils may range from 2,000 to 10,000 kg ha^{-1} or more in the Northeast U.S. Of this, only a small fraction is available and used by trees, perhaps from 10 to 100 kg ha^{-1}. For this reason, total amount of N in the soil may not always be a good indication of nitrogen availability. Amounts of ammonium or nitrate may be measured to determine available forms of N in the soil at any given time. But their production and uptake vary so greatly with temperature, moisture and other conditions that they may not necessarily be a good index of overall availability of nitrogen for any given soil at some given time. In this situation total nitrogen may be the better measurement for plant productivity. Fertilization research has shown that nitrogen deficiencies are not uncommon in agricultural and forest soils in temperate regions and, thus, nitrogen supply appears to be a key factor in controlling the productivity of many soils.

Available N is generally considered to be the **inorganic** form of N; however, not all inorganic N is available [85–88]. Inorganic N in soils is generally considered to be predominantly present in the nitrate (NO_3^-) and exchangeable ammonium (NH_4^+) forms. Nitrite (NO_2^-) is seldom present in detectable amounts, and its determination is normally unwarranted except in neutral to alkaline soils receiving NH_4^+ or NH_4^+-producing fertilizers [85]. Other inorganic N compounds detected in soils include nonexchangeable (mineral-fixed) NH_4^+ (or intercalated NH_4^+), dinitrogen gas (N_2) and "inclusion N_2" (found in rocks or rock-forming minerals), and nitrous oxide (N_2O), as well as some chemical or biological intermediates such as nitric oxide (NO), nitrogen dioxide (NO_2), hydroxylamine (NH_2OH), hyponitrous acid (HON=NOH), and azide (N_3^-).

Until the late 1950s, the combined total of inorganic N forms was thought to constitute less than one to a few percent at most of the total soil N. Later research has shown that many soils contain appreciable amounts of N in the form of fixed NH_4^+ or its organic form, particularly in lower horizons. Unfortunately, many textbooks on soils still perpetuate the erroneous concept that practically all the inorganic N in soils occurs as NO_3^- or exchangeable NH_4^+. But concentrations of mineral-fixed NH_4^+-N have been shown to be as high as 10% of the total N in surface soils with 2:1 type clays and up to 30% in subsoils. Remember though that this form of N is considered relatively unavailable from the perspective of plant production and that NO_3^- and NH_4^+ are still the predominant inorganic forms of N for plant assimilation (or microbial immobilization).

6.5.1 Extraction Methods for Inorganic Soil Nitrogen

Ammonium is held in an exchangeable form in soils in the same manner as exchangeable metallic cations. Fixed or nonexchangeable NH_4^+ can make up a significant portion of soil N; however, fixed NH_4^+ is defined as the NH_4^+ in soil that cannot be replaced by a neutral potassium salt solution [86]. Exchangeable NH_4^+ is commonly extracted by shaking (or perfusing) with 2.0 mol L^{-1} KCl. Nitrate is water soluble and hence can also be extracted by the same 2.0 mol L^{-1} KCl extract. And as mentioned earlier, nitrite is seldom present in detectable amounts in soil and therefore is usually not determined.

The extraction of NO_3^- and NH_4^+ is carried out on air-dry soil for most soil test procedures; however, air drying may result in large changes in soil NO_3^- and NH_4^+ concentrations. Extracting soils in their field-moist condition, immediately after sampling, is the ideal situation; however, this may cause problems with respect to storage and in obtaining a representative subsample. The use of moist soils would be preferred on samples related to biological studies [89]. For most test procedures and fertilizer recommendations, samples air-dried at low temperature (e.g., room temperature) in an NH_4^+-free environment are used (adsorption of gaseous NH_3 may occur).

Analytical procedures

The methods of determination for NO_3^- and NH_4^+ are even more diverse than the methods of extraction [86]. These range from specific ion electrode to manual colorimetric techniques, microdiffusion, steam distillation, and flow injection analysis. Steam distillation is still a preferred method when using ^{15}N; however, for routine analysis automated colorimetric techniques are preferred. They are rapid, free from most soil interferences, and very sensitive. In addition, incubation methods (both aerobic and anaerobic) have been used to determine the potential mineralization rate (organic to inorganic) of N.

A note about units

It is important to clearly specify the units of complex ions such as ammonium, nitrate, and phosphate. Ecologists generally are interested in nitrogen as a nutrient element, so they express concentrations as "mg of nitrogen in the nitrate form per liter" (mg $NO_3^- - N \ L^{-1}$) for solutions, or "mg $NO_3^- - N$ per kg" (mg $NO_3^- - N \ kg^{-1}$) for plant tissue or soil. Regulatory agencies are more often interested in the hazards associated with the ion as a whole, so they use "mg NO_3^- L^{-1}". Geochemists studying charge balance in natural water bodies generally prefer "mmol NO_3^- L^{-1}", which is mg $NO_3^- - N \ L^{-1}$ divided by 14 (the atomic weight of nitrogen), or mg NO_3^- L^{-1} divided by 62, the mass of the nitrate ion.

Note also that **ppm** (parts per million) can mean both mg L^{-1} and mg kg^{-1}. It is a convenient shorthand expression for informal discussion, but it is ambiguous. When doing calculations or formally reporting results, use the full expression, either mg L^{-1} or mg kg^{-1}.

6.5.1.1   *Single Extraction Procedure: Extractable Inorganic Nitrogen*

Purpose

Measure extractable inorganic nitrogen (ammonium (NH_4^+), nitrate (NO_3^-), nitrite (NO_2^-)) in a KCl soil extract using a mechanical vacuum extractor

Materials needed for procedure

1. Potassium chloride, 2 mol L^{-1} (150 g KCl L^{-1}),
2. Cellulose packing, washed in 1 mol L^{-1} HCl and rinsed with distilled water, and
3. Mechanical vacuum extractor, with necessary syringe tubes. "Middle" tubes should be packed with cellulose.

Procedure process

Note: This is the **single extraction** procedure. Depending on your application, you may wish to carry out the **double extraction** procedure.
1. Begin with the extractor cranked all the way down; be sure the locking pin in the top is engaged.
2. Prepare a set of middle syringe tubes (the ones without the short length of rubber tubing) with washed and rinsed packing.
3. Weigh sample into middle syringe (approximately 4 g dry weight equivalent for organic, 8 g for mineral soils). Record weight to 0.01 g. If you are analyzing for NH_4^+/NO_3^-, refrigerate immediately until all tubes are prepared. You should normally leave two tubes empty (no sample, just packing) as blanks, plus whatever spiked (addition of a known amount of material being analyzed for) and duplicate samples (QA/QC checks) (Appendix B) you want to have.
4. Place middle syringes on extractor. Use rubber bands to hold them in place. Place bottom syringe (with plunger and short rubber tube) on extractor. Pull plunger out slightly, and hook rim of plunger under bottom plate of extractor. Push plunger all the way in. Connect rubber tube to middle syringe.

5. Add about 10 mL of extracting solution to middle syringe (or whatever it takes to cover sample). Stir to break up clumps and air bubbles. Sample should be completely wetted.
6. Using a squirt bottle containing extractant, rinse off stirrer (into syringe) and sides of syringe. Try to use no more than about 5 mL.
7. Add any spikes to middle syringe.
8. Put on top syringe. Add 35–40 mL of extracting solution to top syringe. This amount plus the 10–15 mL remaining in the middle syringe should add up to about 50–55 mL, leaving a little extra space to get all the solution down into the bottom syringe.
9. Cover top syringes with paper towels to keep out the dust. Be sure the locking pin on top of extractor is in its hole.
10. Set extractor for 2–3 hours; start motor. Put 50–60 mL of extractant into a bottle labeled with date and "Reagent blank."
11. While extractor is running, label and weigh a set of 60-mL bottles with date of run, ID number, and whatever other information you want. The box that the bottles are stored in should have the complete information, including sample description, date of collection, extracting solution, etc. This information should also go on the data sheet, including the empty bottle weight.
12. At end of run turn off motor. Crank down extractor a few turns to relieve tension on plunger.
13. Carefully withdraw plunger to its limit to pull any remaining solution from middle syringe into bottom syringe.
14. Remove bottom syringe.
15. Carefully squirt extract into labeled and weighed bottle. Record weight of bottle + extract, to 0.01 g. Volume of extract equals mass of extract divided by density. Density of 2 mol L^{-1} KCl is 1.091 g mL^{-1}. Save extract for later analysis (refrigerate for storage longer than one hour).
16. **Clean up!** Wipe up any spilled extract from all extractor surfaces. Salt solutions will corrode the metal. Remove soil and packing from middle syringes using a stiff wire. Dispose of imported soil in proper container. Rinse soil into a soil trap—**no soil in sinks!!** Wash middle syringes in detergent before acid rinse. All syringes and plungers should be rinsed in tap water, then acid rinse (1 mol L^{-1} HCl), then in distilled water; lay them on paper towels to dry.

6.5.1.2   Double Extraction Procedure: Mechanical Vacuum Extractor

1. Begin with the extractor cranked all the way down; be sure the locking pin in the top is engaged.

2. Prepare a set of middle syringe tubes (the ones without the short length of rubber tubing) with washed and rinsed packing.

3. Weigh sample into middle syringe (usually 5–10 g). If you are analyzing for NH_4^+/NO_3^-, refrigerate immediately until all tubes are prepared. You should normally leave two tubes empty (no sample, just packing) as blanks, plus whatever spiked and duplicate samples you want to have.

4. Place middle syringes on extractor. Use rubber bands to hold them in place.

5. Place bottom syringe (with plunger and short rubber tube) on extractor. Pull plunger out slightly, and hook rim of plunger under bottom plate of extractor. Push plunger all the way in. Connect rubber tube to middle syringe.

6. Add about 10 mL of extracting solution to middle syringe (or whatever it takes to cover sample). Stir to break up clumps and air bubbles. Sample should be completely wetted.

7. Using a squirt bottle containing extractant, rinse off stirrer (into syringe) and sides of syringe. Try to use no more than about 5 mL.

8. Add any spikes to middle syringe.

9. Put on top syringe. Add ∼60 mL of extracting solution, so that at the end of the first extraction (60–65 mL) there will still be enough solution in the middle syringe to cover sample (10–15 mL).

10. Cover top syringes with paper towels to keep out dust. Be sure locking pin on top of extractor is in its hole.

11. Set extractor for 2–3 hours; run. Put 50–60 mL of extractant into a bottle labeled with date and "Reagent blank."

12. While the extractor is running, label a set of 125-mL bottles with the date of run, ID number, and whatever other information you want. The box that the bottles are stored in should contain complete information, including sample description, date of collection, extracting solution, etc. This information should also go on the data sheet.

13. Weigh each bottle. Record on data sheet under "Empty bottle, g".

14. Turn off extractor at end of run. Turn crank at top down (clockwise) about two turns to release tension on bottom syringe plungers.

15. Place pinch clamp on rubber tube between bottom and middle syringes. **Carefully** remove lower syringe from tubing, leaving tubing and clamp attached to middle syringe (so residual solution won't drip out).

16. Squirt extract into labeled and weighed bottle. Pull plunger out to the 3 or 4 mL mark, so it can fit over the bottom plate for the second extraction.

17. Carefully reattach bottom syringe to tubing and remove. Check for leaks before moving on to next position.

18. When all bottom syringes have been emptied and replaced, each must be moved outward from its slot. It will be at about a 10° angle from the

vertical, suspended only by the rubber tubing. Be sure tubing is securely attached to both syringe tips (middle and bottom). You can then crank the upper assembly all the way down and slip the bottom syringe plungers under the bottom plate.

19. Add 35–40 mL of extracting solution to top syringe. This amount plus the 10–15 mL remaining in the middle syringe should add up to about 50–55 mL, leaving a little extra space to get all the solution down into the bottom syringe.

20. Set extractor for 2–3 h; cover top and run. Put an additional 50–60 mL of extractant into "Reagent blank" bottle.

21. Turn off extractor at end of run. Turn crank at top down (clockwise) about two turns to release tension on bottom syringe plungers.

22. **Very carefully** pull plunger down by hand, to get last few drops of extract out. Remove bottom syringe (including rubber tubing); squirt extract into appropriate bottle. The bottle will be almost full by the end, and it will overflow if you squirt the syringe too hard.

23. Weigh bottle. Record on data sheet under "Full bottle, g". Volume of extract equals mass of extract divided by density.

24. **Clean up!** Wipe up any spilled extract from all extractor surfaces. Salt solutions will corrode the metal. Remove soil and packing from middle syringes using a stiff wire. Dispose of imported soil in proper container. Rinse soil into a soil trap—**no soil in sinks!** Wash middle syringes in detergent before acid rinse. All syringes and plungers should be rinsed in tap water, then rinse acid (1 mol L^{-1} HCl), then in distilled water; lay them on paper towels to dry.

6.6   Soil Phosphorus

After nitrogen, potassium and phosphorus (P) are the most critical essential elements that influence how plants grow and their productivity levels throughout the world. Phosphorus is especially more limiting to plant growth in the wet tropics where the soils are highly leached. In contrast to nitrogen, phosphorus is not supplied to plants through a process of biochemical fixation so other sources must provide plant requirements for P. Phosphorus is an essential nutrient needed by growing plants where it is a component of adenosine diphosphate (ADP) and adenosine triphosphate (ATP)—two compounds involved in energy transformations in plants. Phosphorus is also an essential component of deoxyribonucleic acid (DNA) and the various forms of ribonucleic acid (RNA) that are needed for protein synthesis. [16,90–92].

Phosphorous is generally taken up by plants in its water-soluble orthophosphate ion (PO_4^{3-}) form. This ion can exist in a number of forms, depending on the pH of the soil [81]:

H_3PO_4 Phosphoric acid; significant only at pH < 3

$H_2PO_4^-$ Predominant form in acid soil solutions

HPO_4^{2-} Predominant form in basic soil solutions

PO_4^{3-} Significant only at pH > 12

As soluble P is removed from soil solution, it is replenished by dissolution of other, more or less labile forms of phosphorus [93]. These include organic P and various iron, aluminum, and calcium phosphate-containing minerals.

Historically, humans have used natural materials—e.g., manures, vegetable material, and bones—as a source of fertilizers and phosphorus supplies for agriculture. These natural materials were used even though the scientific explanation for why they worked was not developed until 1840. At this time, a German chemist named Liebig suggested that dissolving bones in sulfuric acid made the P more available to plants. At about the same time, John Gennet Lawes began experimenting on using a form of acid-treated phosphate or bones as a fertilizer. Unfortunately, the supply of bones was not sufficient and could not keep up with the popular demand for this fertilizer. This triggered a search for other sources of phosphate. In 1842, Lawes was granted a patent to produce phosphate fertilizer using phosphate nodules (erroneously called *coprolites*—fossilized animal excrement). Although Lawes used phosphate nodules collected in Spain for some of his experiments, the first commercial production of phosphate **rock** did not occur until 1847. At this time the mining of "coprolites" began in earnest in Suffolk, Great Britain. Mining of coprolites, however, reached a peak in 1876 when about 250,000 metric tons was mined. Today, the world's supply of P comes mostly from mineral deposits so it is a nonrenewable natural resource. Phosphate rock is produced from three distinct and different forms of deposits: **guano** or guano-derived deposits; igneous or **apatite** deposits; and sedimentary deposits called **phosphorites**. Guano or guano-derived deposits now account for about 5% of the phosphate rock produced in the world with igneous apatites accounting for another 20%. Sedimentary phosphorite deposits account for the majority of the phosphate rock production globally—it is about 75% of the total supplies annually harvested.

Sources of Soil Phosphorus

Phosphorus (P) makes up ~0.12% of the earth's crust. It is present in all soils and rocks, in water, and in plant and animal remains; and it forms complex compounds with a wide variety of elements—about 150 minerals are known that contain at least a minimum of 0.4% P (1% P_2O_5). The highest source of all minable deposits of phosphorus is found as one of the minerals in the apatite group—$Ca_{10}(PO_4,CO_3)_6(F,OH)_{2-3}$. A small percentage of phosphorus is also mined from secondary aluminum phosphate deposits, where the phosphate mineral originally was derived from apatite released during weathering.

Phosphate deposits can be grouped into three broad classes based upon their mineral assemblages. These are:

- Ca phosphates: e.g., fluorapatite, $Ca_{10}(PO_4)_6F_2$;
- Ca-Fe-Al phosphates: e.g., crandallite, $CaAl_3(PO_4)_2(OH)_5 \cdot H_2O$; and millisite, $(Na,K)CaAl_6(PO_4)_4(OH)_9 \cdot 3H_2O$; and
- Fe-Al phosphates: e.g., wavellite, $Al_3(PO_4)_2(OH)_3 \cdot 5H_2O$; variscite, $AlPO_4 \cdot 2H_2O$; and strengite, $FePO_4 \cdot 2H_2O$.

These three classes form a natural weathering sequence in which the **stable Fe-Al phosphates** represent the **final stage of weathering**.

6.6.1 Methodology for Measuring Soil Phosphorus

Methods to determine soil P and its various forms and availability to plants have stimulated a greater understanding of the principles and our scientific knowledge regarding the nature and behavior of P in soils. These methods were designed to characterize the P in the soil system. Many methods exist today, and they vary in principle and technical detail. The selection of a suitable method demands a clear statement of objectives for why the soil P measurement is being taken. Other considerations include soil properties, sample condition or environment, accuracy and reproducibility, and the facilities, equipment, and personnel available to conduct the analyses.

Soil P determinations generally have two distinct phases:

- the preparation of a solution containing the soil P or P fraction, and
- the quantitative determination of the P in this solution.

Quantification of P in most methods is accomplished by colorimetric analysis. The choice of a colorimetric method for determining P typically depends on:

- the P concentration in the solution,
- the concentration of any interfering substances in the solution to be analyzed, and
- the particular acid system involved in the analytical procedure.

The molybdenum-blue methods are the most sensitive (detection limit < 0.001 mg P L^{-1}) approaches developed to analyze soil phosphorus. Therefore, they are the methods that are the most widely used for soil extracts that contain very low concentrations of P or where there is an objective to analyze total P in solution containing orthophosphate ions. In these methods, a phosphomolybdate complex is formed which then turns blue when reduced by ascorbic acid, $SnCl_2$, or other reducing agents. The intensity of the blue color increases with increasing ortho-phosphate concentration but is affected also by other factors such as fluoride concentration [30], acidity, arsenates, silicates, and substances that influence the oxidation-reduction conditions of the system.

Orthophosphate ions may also be determined by **ion chromatography (IC)**, which separates different ions based on their relative mobility through a column containing ion-exchange resins. The IC detection limits are about 0.01 mg P L^{-1}.

Phosphorus in solution may also be determined by **inductively coupled plasma spectroscopy (ICP)**. However, the ICP detects **all** P in solution, not just orthophosphate ions, so any dissolved organic forms would be included. The ICP detection limits are much higher than those of the other methods, about 0.2 mg P L^{-1}.

Analysis of soil P are typically grouped into four broad categories of phosphorus forms:

– Total phosphorus,
– Organic phosphorus,
– Fractionation of soil phosphorus, and
– Phosphorus availability indices.

Total phosphorus

Total P analysis of soils requires the conversion of insoluble materials to soluble forms suitable for colorimetric procedures. The two most widely used methods for extracting the Total P from soils are digestion with $HClO_4$ and fusion with Na_2CO_3. The recommended procedure involves Na_2CO_3 fusion followed by a colorimetric P determination in the water-soluble extract. The method using $HClO_4$was found to give low results with strongly weathered materials and with samples that contained apatite inclusions within constituent primary minerals. However, according to Jackson [30], results will be similar with either method for **most** soils and rocks.

Organic phosphorus

Indirect procedures are used to estimate the total organic P content of soils and are commonly referred to as either extraction or ignition methods. Extraction methods (Eq. (6.14)) involve treating soils with acids, bases, or both, followed by determination of orthophosphate (or oP, or PO_4^{3-}) in the extract before and after oxidation of organic matter, or

$$\text{Organic P} = \text{Total oP in extract} - \text{Inorganic oP in extract} \qquad (6.14)$$

Ignition methods (Eq. (6.15)) utilize either low- or high-temperature ashing to oxidize soil organic matter before acid extraction of total oP. A non-ignited sample is concurrently extracted with acid to determine inorganic P, or

$$\text{Organic P} = \text{oP extracted from ignited sample}$$
$$-\text{oP extracted from non-ignited sample} \qquad (6.15)$$

Essentially all methods currently used for determination of organic P in soils involve the above two approaches. The extraction method involves sequential treatment of a soil with concentrated HCl, NaOH at room temperature, and NaOH at 90°C. Inorganic P is determined immediately after extraction, and total P is determined after oxidation of extracted organic matter with $HClO_4$. Organic P is estimated from the 0.5 mol L^{-1} H_2SO_4 extractable P in a soil sample ignited at 550°C and a non-ignited sample.

The ignition procedure is much simpler, less laborious, and more precise than extraction methods for estimating organic P in soils. However, it is subject to a greater number of errors than extraction procedures, resulting in either an underestimation or overestimation of organic P depending on the properties of the soil analyzed. For many soils comparable organic P values are obtained by using both ignition and extraction methods.

Fractionation of soil P

Fractionation schemes for inorganic P are based on the ability of certain reagents to solubilize phosphates that are present in minerals such as apatite $[Ca_{10}(PO_4, CO_3)_6(FOH)_{2-3}]$, strengite $[FePO_4 \cdot 2H_2O]$, and variscite $[AlPO_4 \cdot 2H_2O]$. This fractionation of soil P is obtained with selective chemical solvents and separated into three discrete classes of compounds, i.e., Al, Fe, and Ca phosphates, some of which could be occluded within coatings of Fe oxides and hydrated oxides. One common extraction procedure involves the sequential P extraction from a soil with:

1. 0.1 mol L^{-1} NaOH to remove non-occluded Al- and Fe-bound P;
2. 1 mol L^{-1} NaCl and citrate-bicarbonate (CB) to remove P sorbed by carbonates during the preceding NaOH extraction;
3. citrate-dithionite-bicarbonate (CDB) to remove P occluded within Fe oxides and hydrous oxides, and
4. 1 mol L^{-1} HCl to remove Ca-bound P.

P availability indices

Phosphorus availability needs to be defined with respect to an external sink, i.e., a plant, plant community, or crop. Plants differ in their ability to extract P from soils due to differences in rooting systems, mycorrhizal associations (see Section 8.3.1), and growth rates. Since any "immediately available" pool of P is constantly replenished through reactions of dissolution or desorption of "less available" P, and through the mineralization of organic P, the pool size of "total available" P is strongly time dependent.

Available P is generally correlated with soluble P or even extractable P (soluble + exchangeable P). In some cases soluble precipitated forms of P may even be included in this correlation. Thus an index of "available" P may be derived by methods that extract exchangeable P and/or soluble P. Extractions may be achieved with solutions of moderately lowered or raised pH, which release P associated with the soil mineral phase, without significantly solubilizing minerals. Alternatively, or in combination with these pH changes, specific anions are introduced that bring P into solution by competing with P sorption sites or by lowering the solubility of cations that bind P in the soil. Based on these principles, numerous extraction methods exist, all of which have some merits and limitations and are used in various parts of the world. The value of each method is based upon and relies on long-term **correlation studies** that establish the relationship between extract and crop response.

Worldwide, the most common methods used to assess P availability are probably the alkaline bicarbonate extract of Olsen *et al.* [94] and the acid ammonium fluoride extract of Bray and Kurtz [95], or various modifications of both of these methods. In addition, an extraction using lactate [96] is popular in Europe. A rationale for the use of bicarbonate or lactate for the extraction of available P is provided by the consideration that plant roots produce CO_2 which forms bicarbonate in the soil solution, and also produce various organic acids similar to lactate that may solubilize soil P. It is hoped therefore that these extractants somehow **simulate the action of plant roots** and thus give a more appropriate measure of plant-available P.

The most common methods used to analyze plant available P are summarized below:

–   The **sodium bicarbonate** extract [94] has been used successfully on a wide range of acid to alkaline soils. Available P is extracted with a solution of sodium bicarbonate at a pH of 8.5 for 30 min. Interference from organic matter dissolved in the solution has frequently been eliminated by sorbing the organic matter onto activated acid-washed charcoal (carbon black) added to the extract, but it is difficult to obtain P-free charcoal. An alternative was therefore developed which eliminates organic interference using polyacrylamide [97]. When the blue phospho-molybdate complex is measured at a wavelength of 712 nm, color interference from the yellow organic matter is negligible. An Olsen test value of 10 mg P per kg is considered to be in the high category [93].

–   The **dilute acid ammonium fluoride** extract (HCl and NH_4F) [95] has been widely used on acid and neutral soils, and a large data base exists. The relatively low acid strength and importance of acidity for the extraction mechanism make the method unsuitable for calcareous or strongly alkaline soils which would partially neutralize the acidity and eliminate the standard test conditions. The exotic composition makes this a purely chemical test that cannot be interpreted in terms of plant function like the bicarbonate or some of the organic acid or chelating extracts. The combination of HCl and NH_4F is designed to remove easily acid-soluble P forms, largely calcium phosphates, and a portion of the aluminum and iron phosphates. The NH_4F dissolves aluminum and iron phosphates by forming fluoride ion complexes with these metal ions in acid solution. A Bray test value of 30 mg P per kg is considered to be in the high category [93].

–   The dilute 0.05M **hydrochloric acid** and 0.0125 mol L^{-1} **sulfuric acid** extract [98]; also called **Mehlich's, Carolina**, or **double acid** extract uses mixed acids unlike the single dilute acid-fluoride method. The mixed acids are more effective than HCl alone as a P extractant on these soils, and the results correlate better with plant response. Relatively greater amounts of Fe phosphate dissolve in the mixed acid solution than in HCl alone. A Mehlich test value of 20–30 mg P per kg is considered to be in the high category, although adjustments for different soil textures may be necessary [92].

–   Another relatively new method (**ammonium bicarbonate**) includes extraction of soil P with 1 mol L^{-1} NH_4HCO_3 and 0.005 mol L^{-1} diethylene triamine pentaacetic acid (**DTPA**) at pH of 7.6. The DTPA is used for the chelation of micronutrients (Zn, Fe, Cu and Mn); the NH_4^+ is used for the extraction of K, the HCO_3^- is used for the extraction of P; and the water base is ideal for NO_3^- extraction.

–   The extraction of soluble P using **water** as described by Olsen and Sommers [90] is useful in measuring the P concentration in water or dilute salt (i.e., 0.01 mol L^{-1} $CaCl_2$) extracts of the soil and in displaced soil solutions and saturation extracts of soil. This index of P availability is used mostly to determine the P concentration level in the soil extract that limits the growth of plants. In soil testing practices, the water or dilute salt extracts represent an attempt to approximate the soil solution P concentration.

–   Using **anion resins** to remove P from soils without chemical alterations or pH changes, is another useful method to index available P. Amounts of P desorbed by resin have been tested as a measure of available P in soils (e.g., Dalal and Hallsworth [99]). This may be accomplished using anion-exchange resins or even by a simpler method using Teflon®-based anion-exchange membranes, which can be cut into strips and used repeatedly and easily [100, 101]. Resins have also been used to assess the availability of residual phosphates (e.g., Bowman and Olsen [102]) and the rate of P release from soils (e.g., Gunary and Sutton [103]) and the buffer capacity of soils (e.g., Olsen and Watanabe [104]).

– The last method for determining P availability is the **isotopic dilution with phosphorus-32**. The available P pool measured by resin extraction is very similar to that assessed with isotopic dilution with ^{32}P [105]. A portion of the solid-phase P in soils exists in equilibrium with the solution-phase P. The amounts present in soils in each phase may be measured by a method based on the isotopic dilution of the radioactive tracer with the stable isotope of the solid and solution phases. The term *isotopic dilution* as used here includes both isotopic exchange involving the solid phase and isotopic dilution in the solution phase. Addition of ^{32}P as orthophosphate ions to a soil-water system causes the following reaction (Eq. (6.16)) to proceed:

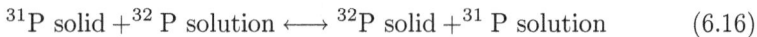

$$^{31}\text{P solid} + {}^{32}\text{P solution} \longleftrightarrow {}^{32}\text{P solid} + {}^{31}\text{P solution} \qquad (6.16)$$

At equilibrium the ratio of the two isotopes in a portion of the solid phase is equal to their ratio in the solution, and the equilibrium constant equals 1 (e.g., Talibudeen [106]). Therefore, the ^{31}P solid can be calculated from Equation (6.17):

$$^{31}\text{P solid} = ({}^{32}\text{P solid}/{}^{32}\text{P solution}) \times {}^{31}\text{P solution} \qquad (6.17)$$

The total P in the system that undergoes isotopic dilution has been termed labile ^{31}P, or from Equation (6.17) it equals the sum of ^{31}P solid and ^{31}P solution [107].

Since available P is a functional concept rather than a measurable quantity, no simple direct measurements are available. However, depending upon the question being asked or the problem being solved, one can choose a method of analysis which may more closely represent the "available P" pool that the plants (or microbes) perceive under the given soil conditions.

Caution: Even small deviations from recommended procedures without careful evaluation may cause serious errors.

A note about units

It is important to clearly specify the units of complex ions such as phosphate, ammonium, and nitrate. Ecologists generally are interested in phosphorus as a nutrient element, so they express results as "mg of phosphorus in the orthophosphate form per liter" (mg PO_4^{3-}–P L^{-1}). Regulatory agencies are more often interested in the hazards associated with the ion as a whole, so they use "mg PO_4^{3-} L^{-1}". Geochemists studying charge balance in natural water bodies generally prefer "mmol PO_4^{3-} L^{-1}", which is mg PO_4^{3-} –P L^{-1} divided by 30.97 (the atomic weight of phosphorus).

6.6.2 Procedure: Extractable Inorganic Phosphorus

Purpose

To estimate "plant-available" phosphorus, using an extracting solution appropriate for a given soil.

Note: Strongly colored extracts may be decolorized by adding 100–200 mg activated carbon (washed in extracting solution) to the sample before extracting. Alternatively, the background color can be subtracted out during autoanalyzer operation.

Bray P-1 Method

Extracts acid-soluble P (mainly calcium phosphates) and a portion of Fe and Al phosphates. Widely used in acid soils. Not appropriate for calcareous soils [95, 108].

Materials needed for procedure

1. Acid ammonium flouride extracting solution (0.03 mol L^{-1} NH$_4$F in 0.025 mol L^{-1} HCl). Dissolve 2.22 g of ammmonium fluoride and 50 mL of 1 mol L^{-1} HCl in DDW; make up to 2 L. Note: NH$_4$ F is toxic; avoid skin contact; waste solution cannot go into sink.
2. Centrifuge tubes, 50 mL.

Procedure process

1. Weigh 1 g of soil into a 50-mL centrifuge tube. Record weight to 0.01 g.
2. Add 10.0 mL of Bray P-1 extracting solution.
3. Shake 5 minutes.
4. Centrifuge; pour off supernatant, filter if necessary. Save extract for analysis on spectrophotometer or flow analyzer. Note: Use 5% boric acid in the flow analyzer carrier solution, to remove flouride interference [30].

Mehlich-1 Method

Also called Carolina or double-acid method. Developed for soils of Piedmont and Mountain regions of North Carolina, which fix P strongly. Useful for acid soils, as it dissolves some iron phosphates. See Nelson *et al.* [98].

Materials needed for procedure

1. Extracting solution (0.0125 mol L^{-1} H$_2$SO$_4$+0.05 mol L^{-1} HCl): 1.4 mL conc. H$_2$SO$_4$+4.2 mL conc. HCl.
2. Centrifuge tubes, 50 mL.

Procedure process

1. Weigh 4 g of soil into a 50-mL centrifuge tube.
2. Add 20 mL of extracting solution and shake for 5 minutes.
3. Centrifuge at 2,000 rpm for 5 min.
4. Pour off and save the supernatant solution; filter if necessary.

Olsen Method

For neutral, alkaline, or calcareous soils. The most reactive phosphates of iron, aluminum, and calcium are extracted. See Olsen *et al.* [94].

Materials needed for procedure

1. Sodium bicarbonate extracting solution (0.5 mol L^{-1}, pH 8.5). Dissolve 42.0 g NaHCO$_3$ in 900 mL DDW. Adjust to pH 8.5 with NaOH. Make up to 1 L.
2. Centrifuge tubes, 50 mL.

Procedure process

1. Weigh 2 g of soil into 50-mL centrifuge tubes. Record weight to 0.01 g.
2. Add 20.0 mL of sodium bicarbonate extracting solution.
3. Shake 30 minutes.
4. Centrifuge; pour off supernatant, filter if necessary. Save extract for analysis on spectrophotometer or flow analyzer.

Note: Before bicarbonate samples can be run on the flow analyzer, they must be acidified with 1 part 2.5 mol L^{-1} sulfuric acid to 5 parts sample. Acidify standards and wash solution also.

Molybdenum Blue

– Ascorbic Acid Colorimetric Analysis for Orthophosphate in 0.5 mol L^{-1}
   Sodium Bicarbonate Extract (pH 8.5)

Note: This procedure is for manual analysis of extracts on a spectrophotometer. For autoanalyzer/flow analyzer, use reagents listed in instrument procedure.

Reagents needed for procedure

1. Sulfuric acid, 2.5 mol L^{-1}: Add 141 mL concentrated H_2SO_4 to 800 mL distilled water, dilute to 1 L. Cool.
2. Reagent A: Dissolve 12 g ammonium paramolybdate $[(NH_4)_6Mo_7O_{24} \cdot 4H_2O]$ in 250 mL distilled water. Dissolve 0.2908 g antimony potassium tartrate $(KSbOC_4H_4O_6)$ in 100 mL distilled water. Add both solutions to 1 L of 2.5 mol L^{-1} H_2SO_4, mix, and dilute to 2 liters with distilled water. Store in dark glass bottle.
3. Reagent B: Dissolve 1.056 g ascorbic acid in 200 mL of Reagent A. This solution should not be kept more than 24 hours.

Procedure process

1. Prepare calibration blank and standards in the expected range of concentrations.
2. In a small test tube, mix 1 mL of extract (or standard), 0.1 mL 2.5 mol L^{-1} H_2SO_4, 3 mL distilled water, and 1 mL Reagent B.
3. Allow at least 10 minutes for color development.
4. Read absorbance at 880 nm on spectrophotometer.

6.7 Soil Carbon and Organic Matter

The organic component of mineral soils (soil organic matter, SOM) has a great effect on soil properties. SOM darkens soil, making exposed soil heat up faster in the spring. It enhances the soil's water-holding capacity and helps to bind soil particles into stable aggregates, creating a mixture of larger and smaller pores for aeration and water movement. Due to the colloidal nature of SOM, it adds to the soil's nutrient retention capacity. It provides a reserve of slowly available nutrients for higher plants and an energy source for microbial communities [16]. Finally, SOM is a large and active component of the global carbon cycle, containing three times the carbon that is contained in terrestrial biomass (1,500 Pg C and 550 Pg C for soil and biomass, respectively; 1 Pg $= 10^{15}$ g), and twice the carbon that is contained in the atmosphere (750 Pg C) [109]. Regional agricultural and forest management practices have a significant effect on whether soils serve as a source or sink of carbon [110, 111].

Many soils, particularly those under forest, have a surface layer of organic soil materials (defined as containing > 20% organic carbon, if it is saturated with water for less than 30 days in a normal year [31]), called **forest floor** (FF), or sometimes **litter** (although properly this term refers only to the most recently deposited, relatively undecomposed foliage, twig, etc. on the surface). It is often ambiguous in published writings whether "soil organic matter" includes the forest floor or, as we prefer, is limited to organic matter (OM) in the mineral soil, with FF a separate category or compartment. Be sure to make this clear in your writing.

Soil carbon can be classified in several ways depending upon the purpose of your analysis. Examples of these classifications are:

- organic vs. inorganic
- humic vs. nonhumic
- living vs. nonliving
- readily decomposable or labile vs. recalcitrant

The carbon/organic matter content of soil can be measured and expressed in many ways. Perhaps the simplest is **Loss on Ignition (LOI)**, the proportion of mass lost when a sample is heated at high temperature (approximately 500°C). This loss is generally termed **organic matter**, and includes carbon plus volatile elements associated with organic carbon (nitrogen, hydrogen, oxygen, and sulfur). However, some water tightly bound to clays will also be driven off as temperature rises, as will some inorganic carbonates (depending on the final temperature), so LOI is not restricted to organic matter.

Soil carbon is a more precisely defined entity than soil organic matter. It can exist in **organic** molecules and in **inorganic** carbonates. Soils contain examples of the common biochemical classes of organic compounds, such as proteins, amino acids, and polypeptides; carbohydrates and polysaccharides (including cellulose); and fats and oils. Lignin is a complex polymeric component of wood that decomposes slowly, but most of these **non-humic** molecules are broken down by microbial enzymes fairly quickly; microbial degradation byproducts, along with resistant lignin subunits, then recombine into **humic** substances (such as **humic acid, fulvic acid**, and **humin**) which are generally high molecular weight, complex-structured molecules with many aromatic (carbon-ring) and phenolic (-OH) groups. Humic molecules give soil organic matter its dark color and high CEC, and tend to be relatively more resistant to further decomposition than are non-humic substances.

There are specialized techniques for extracting and quantifying specific organic compounds or fractions. Methods of determining **total carbon** fall into two categories: **dry oxidation** or **combustion**, and **wet oxidation** [112]. In the former type, a soil sample is heated in a stream of oxygen, converting carbon to CO_2, which is then either trapped and weighed (in older instruments) or determined by infrared or gas chromatographic detectors. The combustion products also contain oxidized nitrogen, hydrogen, and sulfur compounds, which can be processed and sent to various detectors. These procedures can't distinguish between organic and inorganic carbon, so a pretreatment for carbonates may be necessary.

Wet oxidation techniques use an oxidizing reagent such as potassium permanganate or acidic potassium dichromate to oxidize the carbon. Evolved CO_2 may be trapped and measured (in this case, again, carbonates are included), or the unreduced reagent may be determined by colorimetry or titration. Complete oxidation requires carefully regulated heating of the solution, but this introduces problems in preventing loss of water during heating and requires special glassware and other equipment. A simple technique relies on the heat generated by mixing concentrated sulfuric acid and potassium dichromate solution to help oxidize a large part (but **not all**) of soil carbon [113]. The unreduced dichromate is then titrated. As the following reaction shows,

$$2Cr_2O_7^{2-} + 3C^0 + 16H^+ \rightleftharpoons 4Cr^{3+} + 3CO_2 + 8H_2O \qquad (6.18)$$

determining C by titration of unreduced dichromate assumes that the carbon has an average oxidation state of 0 (zero). This is true for carbohydrates, but not for hydrocarbons ($e < 0$) and organic acids ($e > 0$). For soil organic matter in general, this assumption seems to be approximately true. See Manahan [114] for a discussion of redox chemistry.

An advantage of this technique is that carbonates do not react with dichromate, as carbonate-C is already fully oxidized. However, oxidizing and reducing substances found in soils may interfere if they are at high concentrations. These include chloride, ferrous iron, and higher oxides of manganese.

For these reasons, Walkley-Black (W-B) carbon is often used as an index to estimate "readily oxidizable" carbon. A multiplication factor may be used to convert to total organic C. A common value is 1.3, meaning that W-B carbon represents $1/1.3=77\%$ of organic C; however, recoveries ranging from 27 to 100 percent have been measured [112].

6.7.1 Dry Combustion Procedure: Total Soil Carbon and Nitrogen

Purpose

To measure total soil carbon and nitrogen using an automated dry combustion analyzer.

Materials needed for procedure

1. LECO CHN analyzer,
2. Calibration standard, and
3. Oven-dry (at 80°C) soil, sieved and/or ground (about 1 g required).

Procedure process

If the instrument is not already operating, go through the standard daily start-up routine.

1. Run sufficient blanks (usually 4–6) to ensure that instrument temperatures and gas flows have equilibrated, and that blank values are set properly.
2. Choose an appropriate standard for your sample type (soil, plant tissue, or pure chemical).
3. Weigh out and analyze four calibration standards. If they are consistent, go through the calibration procedure.
4. Weigh out replicate subsamples of each sample (2 or 3 reps, depending on sample heterogeneity and your experimental needs) and analyze.
5. Analyze two Quality Control (QC) samples (usually, use calibration standards) at least every hour.

6.7.2   Loss on Ignition (LOI) Procedure: Total Soil Organic Matter

Purpose

To measure total soil organic matter by loss on ignition.

Materials needed for procedure

1. Dessicator
2. Muffle furnace
3. Glazed ceramic crucibles
4. Oven-dry (at 105°C) soil (0.5–2 g)

Procedure process

Note: This procedure may be carried out in conjunction with the dry ash/nitric acid digest procedure, in which case less soil may be used (to reduce concentrations of nutrient elements measured on ICP).

1. Place clean crucibles in drying oven at 105°C for at least 2 hours. Cool in a dessicator. Label (with pencil, not a marker) on the bottom. Record weight to 0.0001 g (**tare wt.**).

2. **If starting with oven-dry soil:** Place 0.5 to 2 g in crucible. Record weight to 0.0001 g (**tare + OD soil wt.**). Go to step 4. **If starting with field-moist soil:** Weigh 0.5 to 2 g dry-weight equivalent of moist soil into a crucible. Record weight to 0.0001 g (**tare + moist soil wt.**, can be used to calculate soil moisture content if desired).
3. Dry to constant weight at 105°C (usually overnight is sufficient). Cool in dessicator. Record weight to 0.0001 g (**tare + OD soil wt.**).
4. Place crucibles with oven-dry soil in muffle furnace. Set furnace to 500°C. Turn on exhaust hood. Ash overnight.
5. Turn off furnace. When temperature has dropped to below 200°, place crucibles in a dessicator to cool. Record weight to 0.0001 g (**tare + ash wt.**).

Calculations

Subtract **tare wt.** to get **OD soil wt.** and **ash wt.**

6.7.3 Walkley-Black Procedure: Soil Carbon

Purpose

To measure soil carbon by the Walkley-Black method.

Materials needed for procedure

1. Oven-dry (at 105°C) soil, sieved and/or ground (about 1 g required)
2. 500-mL Erlenmeyer flask
3. Sulfuric acid, concentrated
4. Potassium dichromate, 0.5 mol L^{-1}: 49.04 g (dried at 105°C) per L
5. Phosphoric acid, concentrated
6. o-phenanthroline—ferrous-iron indicator ("Ferroin")
7. Ferrous sulfate heptahydrate, 0.25 mol L^{-1}: 140 g+15 mL conc. sulfuric acid per L
8. Buret

Procedure Process

Note: Chromium is a toxic metal and must be properly disposed of. Pour wastes into a separate container for later disposal.

1. Weigh soil into flask: about 2 g for low OM/subsurface soils, 0.5 g for high OM/surface soils: sample should contain at least 10, but no more than 25 mg organic C. If you know the total (organic) C percentage, calculate the desired sample weight as 1,000/%C (minimum) to 2,500/%C (maximum).

For example, if the soil has 5% C, the sample should be 200–500 mg. Record weight to 0.0001 g.

2. Add 10 mL of dichromate solution and swirl gently to disperse soil.
3. **In a hood:** Rapidly add 20 mL concentrated sulfuric acid. Immediately swirl gently to mix reagents, then swirl more vigorously for a total of 1 minute.
4. Let flask stand on heatproof sheet in hood for 30 minutes.
5. Add 200 mL of DDI water and 10 mL of concentrated phosphoric acid; swirl gently to mix. (Phosphate ions will complex with ferric iron during the titration.)
6. Add 3–4 drops of indicator, and titrate with 0.25 mol L^{-1} ferrous sulfate. As the endpoint is neared, the color changes from light green to deep green to blue. The endpoint is indicated by a sharp change from blue to red or maroon. It may be difficult to see this color in the presence of suspended soil particles (filter before titrating if necessary).
7. Make a blank determination using the same procedure (but no soil) in order to standardize the FeSO$_4$ concentration.

Calculations

1. The concentration of FeSO$_4$ is nominally 0.25 mol L^{-1} (meq mL^{-1}), but should be calculated (Eq. (6.19)) more accurately from the blank value:

$$X = \frac{(10 \text{ mL Cr} \times 1 \text{ meq/mL})}{??\text{mL Fe}_{\text{blank}}} \tag{6.19}$$

where
mL Cr = volume of 0.5 mol L^{-1} potassium dichromate solution;
mL Fe = volume of ferrous sulfate solution.

2. Calculate the mass (in milliequivalents, meq) of reduced chromium (Cr^{3+}) (Eq. (6.20)), which is equivalent to the mass of carbon that was oxidized. Meq Cr^{3+} is the difference between the original meq of oxidized chromium {meq Cr^{6+}} and the meq of ferrous iron added during the titration {meq Fe}. If some other volume of dichromate solution was used, substitute that value for 10 mL:

$$\text{meq Cr}^{3+} = \{\text{meq Cr}^{6+} - \text{meq Fe}^{2+}\}$$

$$\text{meq Cr}^{3+} = \left\{ 10 \text{ mL Cr} \times \frac{1 \text{ meq Cr}}{\text{mL Cr}} \right\} - \left\{ ??\text{mL Fe} \times \frac{X \text{ meq Fe}}{\text{mL Fe}} \right\} \tag{6.20}$$

3. Finally, calculate Walkley-Black carbon, %C (WB) (Eq. (6.21)):

$$\%C_{\text{WB}} = \frac{??\text{meq Cr}^{3+} \times \frac{1 \text{ meq C}(0)}{\text{meq Cr}^{3+}} \times \frac{1 \text{ mmol C}(0)}{4 \text{ meq C}(0)} \times \frac{12 \text{ mg C}(0)}{\text{mmol C}(0)}}{??\text{mg OD soil}} \times 100\%$$

$$\tag{6.21}$$

Calculate a Walkley-Black recovery factor, f, using your %C determined on the CHN analyzer as the Total soil C (Eq. (6.22)):

$$f = \frac{\%C_{WB}}{\%C_{Total}} \tag{6.22}$$

6.8 Selective Dissolution of Iron and Aluminum

Iron is an essential micronutrient needed by plants at levels less than 0.1% and is present in multiple enzymes in plants as well as it is needed for chlorophyll formation for photosynthesis. However aluminum is not an essential nutrient needed by plants [16]. But, when both elements are present in high concentrations they are toxic to plants and microbes. These higher Fe and Al concentrations are especially common in acid soils. Both iron (Fe) and aluminum (Al) are naturally found in very high concentrations in surficial rock as well as in soils globally [109]. Except for silicon (Si), Al is found in the highest amounts in bedrock (69.3 mg g^{-1} in surficial rock) followed by Fe (35.9 mg g^{-1}) at half the concentrations generally found for Al [109]. So it is understandable that the abundant iron oxides are a common characteristic of soils globally.

By monitoring Fe and Al pools and fluxes scientists are able to not only understand how soils develop but they can better understand ecosystem processes that change in response to natural or anthropogenic disturbances [115]. It allows scientists to know what management practices may alleviate nutrient deficiencies for growing plants. Knowledge of the Fe and Al pools and fluxes allows one to even recognize when a soil ecosystem becomes less resilient or productive due to land-use activities.

One important property of Fe and Al is that they both have mobile forms that can also increase the mobility of other soil minerals. This mobility property can then affect other soil processes and the soil's physical, chemical and biological properties. Some examples of factors that influence these two elements and their subsequent impact on soil processes are: soil development or pedogenesis [116, 117]; plant succession (e.g., deciduous tree dominance shifting to conifers that acidify the soil; [118, 119]; land-uses such as clear-cut harvesting of trees [120]; and pollution (e.g., acid rain, heavy metal contamination of soils due to mining [57, 121]).

Fe and Al can definitely affect soil pH. For example, aluminum can strongly influence the soil pH by hydrolyzing the water molecule forming aluminum hydroxides and leaving the H$^+$ active and hence increasing the soil's acidity. Even iron may contribute to changing the pH in those soils that have reduced irons (Fe^{2+}) such as pyrite (FeS$_2$) soils and/or those that are anaerobic. For example, if a wetland soil is drained (converting an anaerobic reduced soil to an aerobic oxidized soil), the reduced iron sulfur will oxidize producing H^+ ions and thus it can significantly reduce the pH of the soil.

One of the most critical changes that occur in soils due to Al mobility is the leaching of basic minerals (e.g., Ca and Mg) from the surface soil horizons and their fixation, occlusion, or precipitation with subsequent changes in soil pH [16]. This can have many different types of effects. For example, phosphorus availability in the soil is strongly controlled by the soil pH. So in the acid soils in tropical forest regions around the world, phosphorus is commonly unavailable and limiting in those soils [122]. These soil changes, e.g., lower pH and losses of base cations, limit plant growth rates because of decreased available plant nutrients [57, 119].

Generally iron and aluminum are relatively insoluble (especially in neutral and alkaline soil pHs) so may not easily move in the soil profile. However if iron and aluminum become more soluble they can more easily move downward in the soil profile with percolating rainwater and thus can contribute to soil weathering and development (pedogenesis). This soluble state can be instigated by an organic chelate (an organic molecule binding with metal ions) [117]. These organic chelates may be derived from: (1) polyphenol leachates from leaves and litter, or (2) microbial products of humic and fulvic acids [123]. In fact chelates can be easily taken up by plants and when bound with iron are commonly used to treat chlorotic plants that are deficient in iron and cannot get sufficient amounts of less soluble irons from the soil. So iron and aluminum are very important in the natural processes of soil development and are strongly impacted by anthropogenic activities.

Plants, animals and microbes need Fe in small concentrations to maintain their growth rates. Therefore, these organisms can be affected by the Al and Fe effects on pH and even by their potential toxic concentrations in the soil.

6.8.1 Extraction Procedure: Organically Complexed Iron and Aluminum

Purpose

To extract organically complexed iron and aluminum from soil, with 0.1 mol L^{-1} sodium pyrophosphate.

Materials needed for procedurre

1. 125-mL poly bottle, with mark at 125-mL level.
2. 0.1 mol/L sodium pyrophosphate (44.6 g $Na_4P_2O_7$ per liter).

3. Superfloc 16, 0.2% solution in distilled-deionized water.
4. Shaker (e.g., reciprocating shaker).

Method

1. Weigh 2.00 g (dry wt. equivalent) soil into a 125-mL poly bottle.
2. Add 100 mL of 0.1 mol L^{-1} sodium pyrophosphate; cap bottle tightly.
3. Shake overnight (12 to 16 hours) on a reciprocating shaker.
4. Add 2 mL of Superfloc solution.
5. Fill bottle to 125-mL mark with 0.1 mol L^{-1} sodium pyrophosphate solution.
6. Cap bottle and shake vigorously for 15 seconds.
7. Allow to settle for at least 3 days (4–6 days typical).
8. Dilute an aliquot of clear supernatant by 50 times with distilled-deionized water for ICP analysis of iron, aluminum, and manganese (if desired).

6.8.2 Extraction Procedure: Non-crystalline Soil Iron and Aluminum Oxides

Purpose

To extract organically complexed iron and aluminum, non-crystalline Fe and Al hydrous oxides, and amorphous aluminosilicates from soil, with dithionite-citrate solution.

Materials needed for procedure

1. 125-mL poly bottle, with mark at 125-mL level.
2. Sodium dithionite ($Na_2S_2O_4$), purified powder (aka sodium hydrosulfite). Note: Sodium dithionite is a very reactive chemical. It may spontaneously ignite if allowed to become moist, even by atmospheric moisture. Dispense it in a fume hood and wear face and hand protective gear.
3. Sodium citrate dihydrate.
4. Superfloc 16, 0.2% solution in distilled-deionized water.

Procedure process

1. Weigh 2.00 g (dry wt. equivalent) soil into a 125-mL poly bottle.
2. Add 1 g of sodium dithionite and 10 g of sodium citrate.
3. Add 60 mL of distilled-deionized water; cap bottle tightly.
4. Shake overnight (12 to 16 hours) on a reciprocating shaker.
5. Add 1 mL of Superfloc solution.

6. Fill bottle to 125-mL mark with distilled-deionized water.
7. Cap bottle and shake vigorously for 15 seconds.
8. Allow to settle for at least 3 days (4–6 days typical).
9. Dilute an aliquot of clear supernatant by 50 times with distilled-deionized water for ICP analysis of iron, aluminum, and manganese (if desired).

Chapter 7

Total Plant and Soil Nutrient Analysis (Digestion)

To determine total concentration of an element in a soil or plant sample, much stronger methods need to be applied than conducting a simple extraction. Digestion is needed in order to break strong covalent organic and (in the case of soil) inorganic bonds and to put the species of interest into solution. While there are techniques for measuring total composition of solid samples (such as X-ray fluorescence and electron micro-probe) they will not be discussed here. Analysis of total carbon and nitrogen is covered in Section 6.7, Soil Carbon and Organic Matter.

A note of caution is in order when using the word "total" in connection with digests, particularly of soil. Very reactive reagents such as hydrofluoric or perchloric acids are required to completely dissolve silicate minerals. The two methods presented in this chapter are less aggressive. Virtually all organically bound fractions of an element will be released, plus an unknown proportion of the mineral fraction. For example, the following table (Tab. 7.1) shows the percent recovery of 11 elements in a modified Kjeldahl digest of National Institute of Standards & Technology (NIST) Standard Reference Material 2711, Montana Soil [124]. Carbon is 2% (non-certified value).

Two digest methods are presented here: a **wet oxidation** (sulfuric acid/block digest, or modified **Kjeldahl**) procedure, and a **dry oxidation** (dry ashing, followed by nitric acid dissolution) technique.

7.1 Wet Oxidation Method

The original Kjeldahl method [125] was developed for the determination of total nitrogen in organic materials, and has been used extensively for N analysis in feeds, plant tissue, and soil (see Bremner [126] for a thorough discussion of the evolution of Kjeldahl methods, their modifications, and pretreatments necessary

Tab. 7.1 Percent recovery of 11 elements in a modified Kjeldahl digest [124].

Element	Certified Value (mg kg^{-1})	Measured Value (mg kg^{-1})	Recovery from Digest (%)
Al	65,300	2,172	3.3
Fe	28,900	2,293	7.9
Na	11,400	1,344	11.8
K	24,500	8,536	34.8
Mg	10,500	5,064	48.2
N	1,900*	1,220	64.2
Ca	28,800	20,445	71.0
P	860	643	74.8
Cu	114	95	82.9
Mn	638	547	85.7
Zn	350	303	86.5

* No certified value for N. Total N measured 1,900 N mg kg^{-1} (or 0.19%) on LECO CHN-600. Low recovery is probably due to lack of catalyst in digest mixture.

for certain analyses). More recently, modified Kjeldahl procedures have been used to determine total concentration of many elements in both plant and (with the caveat mentioned above) soil samples.

Kjeldahl methods use sulfuric acid and high temperature to oxidize organic molecules, converting them to carbon dioxide, water, and dissolved ionic forms of other organically bound elements. Many highly resistant molecules are more quickly and completely broken down by the addition of various modifiers:

– strong oxidants like hydrogen peroxide or potassium permanganate.
– salts (potassium, sodium and/or copper sulfate) to raise the boiling point of the acid, allowing higher temperatures to be reached.
– metallic catalysts, such as mercury and selenium.

Under these conditions, most nitrogen is converted to ammonium ion (NH_4^+), which can be detected in a variety of ways. However, recovery is not complete with compounds containing N—N and N—O linkages (e.g., nitrites, nitrates, azo, nitroso, and nitro compounds, hydrazines, hydrazones, oximes, pyrazolones, isooxazoles, 1,2-diazines, 1,2,3-triazines). Other pretreatments may allow inclusion of such compounds. In addition, clay-fixed NH_4^+ may be locked in the clay lattices during digestion, unless the sample is pre-treated with water to expand or with HF to dissolve the clays. Traditionally, nitrogen determined by these methods is referred to as total Kjeldahl nitrogen (TKN), rather than simply as total N.

Formerly, digestion was followed by a **distillation** step, in which the acid digest was made alkaline, converting ammonium ion to ammonia gas ($NH_4^+ + OH^-$

——→NH_3+H_2O). The ammonia gas was carried over into another container by steam and bubbled through an acidic solution, thus converting it back to ammonium ion. Concentration was determined by titration. This step has been replaced by automated colorimetric techniques that are more sensitive, allowing the use of smaller acid volumes and smaller samples. Block digesters accommodating 40 or 80 tubes have replaced bulky digestion and distillation racks.

Combustion analyzers have largely superseded digestion for total N determination. In order to eliminate the use of hazardous metals such as Se and Hg, an experiment was conducted in the Yale School of Forestry and Environmental Studies laboratory to test the efficiency of digestion without a catalyst (selenium) on recovery of various elements from NIST Standard Reference Material 1515 (Apple Leaves) [124]. Absence of Se had no effect on recoveries of Al, Ca, Cu, Fe, Mg, Mn, P, or K (ranging from 97%–104%). Nitrogen recovery dropped from 98.5% with Se to 95.1% without it. Although this difference was not statistically significant (at 5%) in this experiment, it seems certain from the extensive literature on Kjeldahl methods that a catalyst is necessary for complete nitrogen recovery. In other experiments on soils, nitrogen recovery with the modified Kjeldahl method presented here was 80%–90% of total N (as determined by the LECO CHN analyzer). However, its use for total nutrient analysis for the other elements listed seems valid (again, for **plant** tissue only; total **soil** digestion requires hydrofluoric acid (HF) or a comparable procedure). Note that lead recoveries in sulfuric acid digests are very low due to the precipitation of lead sulfate. When samples have high concentrations of aluminum and iron, digestion recovery of these elements is low, apparently due to their precipitation as insoluble oxides.

The wet oxidation method presented here includes concentrated sulfuric acid, 30% hydrogen peroxide, and lithium sulfate [127] but no catalyst. The H_2SO_4 and H_2O_2 are for oxidizing the organic material to release the organically bound nutrients; Li_2SO_4 raises the boiling point of H_2SO_4 (from 290°C to >350°). The use of Li_2SO_4 allows for analysis of K, Na, and Cu (which would not be possible in methods that use K_2SO_4, Na_2SO_4, or $CuSO_4$ for raising the acid's boiling point).

7.2   Dry Oxidation Method

Nitrogen

In the classical Dumas method [128] of determining N, the sample is heated (over 600°C) with copper oxide in a stream of carbon dioxide. The gases liberated are fed over hot metallic copper to reduce nitrogen oxides (mainly N_2O) to N_2, and then over CuO to convert CO to CO_2. The CO_2 is absorbed by concentrated alkali, and the volume of N_2 gas is measured [125]. Modern automated systems follow the same basic procedure (see McGill and Figueiredo for a review [129]). When combined with a carbon analyzer (as in the LECO CHN-600), combustion occurs in a stream of pure oxygen, to yield CO_2, H_2O, and NO_x. Carbon (as CO_2) and hydrogen (as H_2O) can be determined with infra-red detectors. The combustion gases then enter a reduction furnace containing copper and copper oxide at 650°C, which removes excess O_2 and converts NO_x to N_2. Then CO_2 and H_2O are absorbed in chemical traps and N_2 is measured by a thermal conductivity detector. Quantification normally requires about 4 minutes.

Dumas techniques have the advantage of requiring less laboratory space, do not produce noxious fumes, and include all forms of N without lengthy pre-treatments [130]. For tracer studies, they avoid digestion, distillation, titration, evaporation, and subsequent oxidation of NH_3 to N_2.

Dry ashing with nitric acid dissolution

Dry oxidation (ashing) may also be used to liberate nutrient elements from an organic matrix, prior to putting them into solution. Dry ashing must always be carried out at as low a temperature as possible and the operation cannot be hurried. To obtain a good ash, free from excessive amounts of carbon, requires considerable experience and patience. Some elements may be lost if the ashing is carried out at too high a temperature. High temperatures also favor the formation of complex silicates that are not readily soluble in hydrochloric acid, even after prolonged digestion. It is probable that many of the apparent losses of the inorganic constituents, especially trace metals, during ashing are due to this. Fusion of some of the salts of the ash may occur if the temperature is too high and these fused salts surround particles of unburned carbon, excluding air and preventing their free combustion. If ashing is carried out too quickly, deflagration occurs, producing excessive local heating within the glowing mass. Some analysts prefer to treat the sample with sulfuric acid before ashing so as to obtain a sulfated ash, which is less fusible than ordinary ash. However, this does not prevent the formation of varying amounts of difficultly soluble silicates, particularly those of the trace elements.

No loss of phosphorus occurs during the ashing of most plant materials, provided that the temperature does not exceed 600°C. However, when ashing seeds or other plant materials low in basic constituents, the sample should be thoroughly mixed with calcium or magnesium acetate or nitrate prior to ashing (of course that limits the elements you can analyze for, depending upon which chemicals are added). The acetates are recommended in preference to the nitrates since the strongly oxidizing action of the latter may cause violent deflagration, with considerable rise in temperature of the ash. In the presence of added magnesium salts some research has shown that the temperature may reach 800°C without the loss of phosphorus. Chlorine and sulfur are largely lost during ashing unless a basic substance such as lime or sodium carbonate is intimately incorporated with the material before ashing.

Some methods of ashing are followed by digesting the siliceous residue and salts with HCl. However, no matter how carefully this is done, the residue retains some constituents. In particular, significant amounts of some of the **trace elements** are strongly retained. Even after prolonged digestion with HCl, they can only be recovered by alkaline fusion or by solution of the silica in HF. In many cases more than one-fourth of the total amounts of manganese, copper and zinc have been found in the insoluble residue (the absolute amounts, of course, are not enough to affect the mass of the silica, if the mass of the silica is what is being sought). For this reason a more oxidizing acid (nitric, instead of hydrochloric) can be used for the post-ashing digestion phase. Siccama and Johnson [131] have found that trace elements are efficiently extracted using HNO_3 as the post-ashing digestion mixture, although manganese recoveries from Standard Reference Materials appear to be low [124], probably from the formation of insoluble oxides.

7.3 Total Dissolved Carbon and Nitrogen in Water

Although not strictly a soil property, concentrations of carbon and nitrogen dissolved or suspended in water (from streams, groundwater, lysimeters, or soil extracts) are often of interest to people studying soils. Depending on sample pre-treatment and instrument settings, a number of different species can be distinguished: by filtration, particulate vs. dissolved C and N; by acidifying and purging (to remove carbonate, bicarbonate, and carbon dioxide), organic vs. inorganic carbon.

Dissolved carbon is determined by injecting a small sample into a heated column containing a catalyst in a stream of oxygen. Carbon in the sample is oxidized to carbon dioxide, which is then measured by infrared absorbance [132].

Dissolved inorganic nitrogen (ammonium, nitrate, and nitrite) can be determined directly on an automated colorimetric analyzer (autoanalyzer, flow analyzer, etc.) or an ion chromatograph (IC). Organic nitrogen must be digested to convert it to an inorganic form before analysis. So for the water sample the Total Dissolved Nitrogen = Dissolved inorganic nitrogen + Dissolved Organic nitrogen. A modification of the Kjeldahl digestion procedure can be used to convert organic N to ammonium [132]. Another procedure uses an alkaline potassium persulfate solution and elevated temperature and pressure (in an autoclave) to oxidize organic N (as well as inorganic ammonium) to nitrate [133]. Dissolved organic N is obtained by subtracting inorganic N (measured separately) from total N (in the digest).

7.4 Modified Kjeldahl Digest Procedure: Sulfuric Acid Digest for "Total" Nutrients

Purpose

To measure "total" nutrient content of soil or plant tissue by a sulfuric acid digest (block digester method).

   Note: This is a hazardous procedure, using several highly corrosive and/or toxic chemicals. Do not carry it out without qualified supervision!

Materials needed for procedure

1. Laboratory safety equipment: goggles, lab coat or apron, rubber gloves.
2. Acid digest solution (concentrated sulfuric acid, with 33.33 g lithium sulfate (monohydrate) per liter of acid, **or** 28.62 g anhydrous lithium sulfate per liter of acid).
3. Hydrogen peroxide (30%).
4. Filter paper (Whatman No. 40 (ashless), 5.5 or 7 cm).
5. Block digester, with 100-mL tubes.
6. Teflon boiling chips (optional).

Procedure process

1. Weigh 0.1–0.3 g of oven-dried sieved soil or ground plant tissue onto a folded No. 40 filter paper (smaller amounts give better recovery of iron and aluminum; on the other hand, you may lose the ability to detect K, P, or trace elements). Record weight to 0.0001 g. Wrap the sample tightly and place it into digest tube. Include at least two blanks (filter paper, no sample) and two Standard Reference Materials or spiked samples in every batch of 40.
2. Note: Wear proper protective gear when performing the following steps.
3. Turn on hood fan.
4. Working in a hood, add 2.5 mL of acid digest mixture to each tube. Note: add acid, then peroxide (step 4) to one tube, then move on to the next tube, rather than adding acid to all 40 tubes, then peroxide to all 40 tubes.

5. Keeping mouth of tube pointed back into hood, add 30% hydrogen peroxide with an eyedropper. After each dropperful, swirl tube until foaming subsides. If foaming approaches mouth of tube, place tube in cool water to slow reaction. However, the goal in this step is to allow the reaction to proceed as vigorously as possible without foaming over.

6. Continue adding peroxide until there is little or no reaction (usually 4–8 mL). As much as possible, try to rinse down walls of tube as you add peroxide. For root, forest floor, or soil samples, allow tubes to stand, from several hours to overnight (forest floor with decayed roots needs the longest time)—the pre-digest step.

7. Optional: add 4–6 Teflon boiling chips to each tube, to minimize bumping during heating. Experience with a given type of sample will tell you whether the chips are necessary, but they seem to be less so with the smaller acid volume (2.5 mL) we are currently using.

8. Carefully place rack of tubes into block.

9. Set temperature to 160°C (the minimum dial setting). Cook at this temperature until water begins boiling off. Watch for excessive foaming. A drop of iso-amyl alcohol will reduce foaming briefly; a few drops of hydrogen peroxide will help break down the resistant organic substances producing the foam. In either case, wear a face shield as you add it. If foaming is not a problem, water can be added carefully to rinse down sides of tube at this stage.

10. When most water has boiled away, increase temperature setting. When temperature reaches 250°C, carefully add 1–2 mL of hydrogen peroxide. Dark solution should become clear.

11. Increase temperature to 300°C. Watch carefully to be sure that digest doesn't foam over.

12. Cook at 300°C for 30 minutes. Note: For maximum recovery of iron and aluminum, do not exceed 300°, and do not cook at this temperature for longer than one-half hour, even if complete clearing has not occurred. Watch carefully for excessive foaming and boiling over during this step. If a very thick, semi-solid foam "plug" forms and moves upward, remove that tube. After it cools, add peroxide to break up foam, then place tube back in block. Swirl tubes occasionally to wash down sides.

13. When digestion is complete, turn off heater, and remove tubes. Place on asbestos mat next to block to cool (about 30 minutes). Note: For maximum recovery of iron and aluminum, proceed to the next step as soon as the tubes are cool enough to handle: 25–30 minutes maximum. Do not let the undiluted acid sit overnight.

14. Add DDI water to bring sample close to desired volume (there is some expansion of volume from heating when acid and water mix). The tubes are marked at 100 mL. For a smaller final volume (for lower detection

limits), add about 25 mL of DDI to the tube, then filter sample (with No. 41 ashless paper) into a 50-mL volumetric flask. Be sure to transfer all the liquid from the tube into the filter, using several washes of DDI water.

15. Allow diluted sample to cool to room temperature before making up to final volume.

16. Mix diluted digest: stopper tube and invert 3 times. Rinse stopper before going on to the next tube. Using two stoppers alternately allows you to keep track of where you are. For soil digests, after mixing, let tubes stand to allow any sediment to settle (several hours to overnight). For maximum recovery of iron and aluminum in plant digests, pour into bottle immediately after mixing.

17. Pour diluted digest into poly bottle. If there is sediment on the bottom, pour liquid off carefully so sediment stays in the tube. You only need about 20 mL for ICP analysis. Note: make up ICP calibration standards in 2% digest acid (20 mL per liter).

18. During clean-up, pour remaining digest solution and boiling chips into large ceramic funnel by sink, to recover chips. Digest solution must be neutralized (with baking soda or technical grade sodium hydroxide) before it goes down the drain. Tubes get washed with detergent, and then rinsed with tap water, 1 mol L^{-1} HCl, and three times with distilled water.

An approximate timetable for the block digest procedure

1. Weighing out samples	1 hour
2. Add acid and peroxide	1 hour
3. Predigest	0 hour for soil, to overnight for forest floor (or organic detritus)
4. Digest on block	2–3 hours
5. Cool	0.5 hour
6. Add water	0.5 hour
7. Cool	1 hour
8. Add water to volume	0.5 hour
9. Mix	0.5 hour
10. Let settle	0 hours to overnight
11. Pour into bottles	0.5 hour
12. Clean tubes	1 hour

The total time for the block digestion procedure is 9–11 hours, spread out over two or three days. Plan your schedule accordingly.

7.5 "Total" Nutrient Analysis Procedure: Dry Ashing Followed by Nitric Acid Digest

Purpose

To measure "total" nutrient content of soil or plant tissue by dry ashing followed by nitric acid digest.

Materials needed for procedure

1. Laboratory safety equipment: Goggles, lab coat or apron, rubber gloves
2. Glazed ceramic crucibles, of an appropriate size
3. Muffle furnace
4. Nitric acid, 6 mol L^{-1} (375 mL conc. acid per liter)
5. Hotplate
6. Funnel
7. Filter paper (Whatman No. 41, 11 cm)
8. Volumetric flask, 50 mL

Procedure process

1. Place clean crucibles in drying oven at 105°C for at least 2 hours. Cool in a desiccator. Label (with pencil, not marker) on the bottom. Record weight to 0.0001 g.
2. Include at least one blank (no sample in crucible) and one standard reference material in each batch. A batch consists of 10 to 20 samples, which can conveniently fit into the furnace (depending on the size of the crucible) or can fit into an available time frame.
3. If starting with oven-dry soil: Place 0.3–0.5 g of soil or 0.2–0.3 g plant material in crucible. Record weight to 0.0001 g. Go to step 5. If starting with field-moist soil: Weigh 0.3–0.5 g dry-weight equivalent of moist soil into a crucible. Record weight to 0.0001 g.
4. Dry to constant weight at 105°C (usually overnight is sufficient). Cool in desiccator. Record weight to 0.0001 g.
5. Place crucibles with oven-dry sample in muffle furnace. Set furnace to 500°C. Turn on exhaust fan. Heat in furnace overnight.
6. Turn off furnace. When temperature has dropped to about 300°, place crucibles in a desiccator to cool. Record weight to 0.0001 g.
7. Note: Wear proper protective gear when performing the following steps.
8. If you have a large number of samples, it might be best to prepare all the filters with filter paper, 50-mL volumetric flasks, and sample solution bottles at this time for efficient use of time.

9. Add 8 mL of 6 mol L^{-1} HNO_3 to each sample. Place crucible on hotplate at 300°C. Swish periodically to prevent "ring around the collar". Remove crucible just as it starts to boil.
10. Rinse filter paper with 2%–5% nitric acid. Filter sample into a 50-mL volumetric flask. Be sure to transfer all the material from the crucible into the filter, using several washes of DDI water.
11. Wash filter several times with DDI water, making sure that the 50-mL volumetric isn't filled up past the 50-mL mark!
12. Add distilled/deionized water to bring to volume in the 50-mL volumetric flask.
13. Pour diluted digest into poly bottle for ICP/solution analysis. **Note:** make up ICP calibration standards in 5% nitric acid.
14. Cleanup: Wash crucibles in detergent. Rinse funnels, flasks, and crucibles with tap water, dilute acid (1 mol L^{-1} HCl), and three times with distilled/deionized water.

7.6   Total Dissolved Nitrogen in Water Procedure: Persulfate Oxidation

Note: Procedure modified from D'Elia *et al.* [133].

Purpose

To measure total dissolved nitrogen in a water sample.

Materials needed for procedure

1. Autoclave.
2. Known concentration of Dissolved Organic Nitrogen for use as a "spike".
3. Alkaline persulfate oxidizing reagent: 9.0 g NaOH + 20.1 g $K_2S_2O_8$ per L of distilled-deionized water. Make fresh just before use.
4. 1 mol L^{-1} hydrochloric acid: 83 mL concentrated HCl per L of distilled-deionized water.
5. 25 mm × 150 mm borosilicate glass screw-cap test tubes (50-mL capacity).
6. Glycine (NH_2CH_2COOH) standard stock solution, 100 mg N L^{-1}: dissolve 0.0536g glycine in distilled-deionized water acidified with a few drops of concentrated hydrochloric acid; make up to 100 mL. Store at 4°C.
7. Nitrate standard stock solutions: 1000 mg N L^{-1}: dissolve 0.722 g potassium nitrate in distilled-deionized water acidified with a few drops of concentrated hydrochloric acid; make up to 100 mL. Store at 4°C. (You may

also use 0.607 g sodium nitrate.) For 100 mg N L^{-1}, use 1/10 the amount of nitrate salt. Store at 4°C.

Procedure process

1. Add 10.0 mL of sample to test tube. In each batch of samples, include a water blank and a Dissolved Organic Nitrogen (DON) spike (add 10.0 mL DDI water to these two tubes), and an appropriate number (7 is suggested) of tubes for calibration blank and standards (do not add water to these calibration standard tubes at this time).
2. Add 0.100 mL of glycine stock solution (100 mg N L^{-1}) to the DON spike tube; this should give a final concentration of 1.0 mg N L^{-1}.
3. Add 5.00 mL of persulfate oxidizing reagent to each tube (samples, blank, spike, and calibrants). Cap tubes immediately.
4. Place remaining persulfate solution in a flask with a small beaker over the mouth for a cap (not airtight), for the flow analyzer wash solution.
5. Autoclave sealed samples, standards, and wash solution for 90 minutes at 110°C (230°F). Set steam exhaust on SLOW (Liquids).
6. When the pressure in the autoclave has dropped to 0 and the temperature is less than about 50°C, open autoclave; remove tubes and wash solution. **Caution:** hot steam will still be present when you open the autoclave. Allow them to cool to room temperature.
7. Add 10.00 mL DDI water to the calibration standard tubes. Add the appropriate amount of nitrate standard stock solution to make the desired series of calibration standards, using the formula: mL stock = (10 mL × standard concentration) / stock concentration. (See suggested calibration series at end of procedure—Tab. 7.2)

Tab. 7.2 Suggested calibration series for Dissolved Organic Nitrogen (DON) determination.

Standard conc. (mg N L^{-1})	Stock conc. (mg N L^{-1})	mL of stock to make 10 mL of standard
0.0 (Blank)	100	0.000
0.1	100	0.010
0.2	100	0.020
0.5	100	0.050
1.0	1000	0.010
2.0	1000	0.020
5.0	1000	0.050

8. Add 0.4 mL of 1 mol L^{-1} HCl to each tube; add 0.08 mL of 1 mol L^{-1} HCl for each mL of persulfate solution to the bottle of wash solution.

9. Add 2 mL DDI water for each mL of persulfate solution to make the wash solution.

10. Analyze for nitrate-N, using either the flow analyzer or ion chromatograph. Calibrate with the digested standards.

Chapter 8
Soil Biology Characterization

Organisms that live on or in the soil include vascular plant roots, microbes (archaea, bacteria, fungi, algae, and lichens), invertebrates (collembola, mites, insects, nematodes, earthworms, spiders, etc.) and vertebrates (voles, gophers, reptiles, etc.). They are functionally important in soil ecosystems because of their roles in decomposition and therefore in the storage and flux of carbon and nutrients, their influence on soil structure and water holding capacity, they are partners symbiotic relationships, their delivery rates of nutrients needed to maintain plant productivity, and their role as disease agents and vectors.

Soil organisms, excluding roots and vertebrate animals, typically comprise less than 1%–2% of total ecosystem biomass but their importance in ecosystem processes and biodiversity far outweighs their biomass contribution. They are the source of most of the reported ecosystem biodiversity.

Soil biology is important to characterize in any ecosystem since they are the agents that break down senesced plant tissues. Their decomposition activity releases nutrients stored in senesced plant tissues so they continue the cycle for plants to acquire the limiting nutrients they need for growth. Some of the soil organisms have also formed symbiotic associations with plants, e.g., mycorrhizal or nitrogen-fixing organisms, and facilitate plants being able to grow under nutrient limiting conditions or even chemically toxic soils. Plants would not grow on former mine spoils without being inoculated with these symbiotic organisms. These symbiotic associations are also important because bacteria and fungi are able to increase the plants access to nutrients beyond the capacity of their root systems.

The decomposition activities of the soil biota are also the agents responsible for producing the organic matter that accumulates in soils. If the population size and composition of this biota were reduced, e.g., land-use activities, soil health would decrease and plant acquisition of nutrients would decrease. Even though the soil biology is more diverse than all the plant communities combined, their size has made it difficult, if almost impossible, for scientists to study and measure this biodiversity. It is especially difficult to study them in the natural environment without altering their composition and functions since the methods mostly remove them from their habitat. Therefore, researchers have developed many indirect methods (see Section 8.4) to study the functions of these organisms to understand their critical roles in maintaining the structures and functions of soils and therefore ecosystems.

Soil organisms are classified into three domains: Eukarya (eukaryotic organisms), Bacteria (prokaryotic true bacteria), and Archaea (prokaryotic Archaeabacteria—halophiles, methanogens, and extreme thermophiles). The Eukarya include plants, animals, protozoans, fungi, and fungal-like organisms (groups with cellulose in cell walls and slime-molds). Most biologists now use the Tree of Life approach that utilizes DNA analysis to examine phylogenetic relationships among groups of organisms rather than placing them in Kingdoms.

Methods to describe soil biological characteristics and activity are discussed in this section, including soil microbiology (see Section 8.1), soil invertebrates (see Section 8.5), mycorrhizas (see Section 8.3), and indirect indices of soil biological activity (see Section 8.4) (i.e., soil respiration, litter decomposition rates, enzyme activity, functional biodiversity, substrate utilization profiles and molecular tools). Only the most common methods are described in detail. More detail on soil biology characterization can be found in Weaver *et al.* [134], Robertson *et al.* [135], Kirk *et al.* [136], and Schinner *et al.* [137] [1] .

Soil sampling has been described previously and the sample coring and handling techniques used for other analyses are satisfactory to collect samples to describe soil biology. However, a few additional suggestions are needed for sampling and measuring the functions and structures of soil organisms since their biological activity occurs over a time period of a few hours. Samples should be stored in bags that retain moisture but allow respiration. These bags should be made of thin polyethylene and should be placed in a temperature-controlled environment immediately upon collecting a sample in the field. A portable cooler has been found satisfactory for the transportation of samples. Immediate processing, although most desirable, is often not possible. If samples need to be stored before processing, short-term sample storage at 5°C is often used.

8.1 Soil Microbes

The dominant soil microbes are archaea and bacteria, fungi and algae. In this section methods are presented to determine what soil microbes are present as well as their numbers, diversity and biomass.

---

1 Some material was adapted from *Standard Soil Methods for Long-Term Ecological Research* edited by G. Philip Robertson, David C. Coleman, Caroline S. Bledsoe and Phillip Sollins (1999) 1054 words from pp. 193-194, 207, 212-213, 216, 220-221, 246-248, 361, 362, 368, 391-393, 395, 396, 401 and 402 (adapted). By permission of Oxford University Press, USA.

8.1.1 Archaea and Bacteria

Archaea, like bacteria, are prokaryotes and have no cell nucleus or any other organelles within their cells. They have an independent evolutionary history from bacteria. Initially, seen as extremophiles first detected in harsh environments, e.g., hot springs and salt lakes, they are now known to occur in a broad range of habitats, including soils, oceans, and marshlands. Many are mutualistic but none are pathogens. Specific sampling methods for archaea are not discussed here.

Bacteria are prokaryotic single-celled organisms with population sizes ranging from 100 million to 3 billion per gram of soil. Some may be filamentous like the Actinomycetes. Under favorable conditions, they are capable of very rapid reproduction by binary fission (dividing into two). One bacterium is capable of producing 16 million more bacteria in just 24 hours. Most soil bacteria live close to plant roots and are often referred to as rhizobacteria. Bacteria live in soil water, including the film of moisture surrounding soil particles, and some are able to swim by means of flagella. They typically have three basic shapes—cocci (round), rods, and spiral. Bacteria have no specialized structures for dispersion. They reproduce asexually and are dispersed passively in the atmosphere by being blown off plant or soil surfaces, or they can move in water. Some have thick-walled specialized survival spores (endospores) that are particularly resistant to adverse conditions. Some bacteria produce these spores which enable them to survive even in boiling water. Typical cell diameters range from 0.2 to 2 μm and from 1 to 10 μm long for non-spherical bacteria.

The majority of the beneficial soil-dwelling bacteria need oxygen (aerobic bacteria), while those that do not require air are referred to as being anaerobic. Bacteria tend to cause putrefaction of dead organic matter. Aerobic bacteria are most active in a soil that is moist (but not saturated, as this will deprive aerobic bacteria of the air that they require), and has a neutral soil pH. These conditions provide the most available food (carbohydrates and micronutrients from organic matter).

Hostile conditions will not completely kill bacteria; rather, the bacteria will stop growing and move into a dormant stage. Those individuals with pro-adaptive mutations may compete better in the new conditions. Bacteria are Gram positive (purple in Gram stain developed by a Danish microbiologist in 1884) while others are Gram negative (red). Some Gram positive bacteria produce spores in order to wait for more favorable circumstances, and Gram negative bacteria move into a "non-culturable" stage. Bacteria are colonized by persistent viral agents (bacteriophages) which determine gene messaging in bacterial hosts.

Bacteria are important in the process of decomposition, particularly for hazardous chemicals and oil spills, but most are effective at decomposing substrates that are high in cellulose and lignin. They are especially important in soil nitrogen transformations (N fixation, nitrification, denitrification, etc.), methane production and methanogenesis, and sulphur and iron oxidation and reduction.

The numbers of species and genera for the bacterial phyla are presented below in Table 8.1. Only about 5,000 species have been described.

Tab. 8.1 Numbers of species and genera in the bacteria phyla [138].

Phylum	No. Species	No. Genera
Aquificae	27	12
Xenobacteria	29	11
Chrysogenetes	1	1
Thermomicrobia	13	6
Cyanobacteria	78	62
Chlorobia	17	6
Proteobacteria	1,644	366
Firmicutes	2,474	255
Planctomycetes	13	5
Spirochaetes	92	13
Fibrobacter	5	3
Bacteroids	130	20
Flavobacteria	72	15
Sphingobacteria	76	22
Fusobacteria	29	6
Verrucomicrobia	5	2

The most numerous bacterial phyla Proteobacteria and Firmicutes are common in all soils. Proteobacteria are gram negative, flagellated, round or oval, or rod shaped. Typical examples include the enteric bacterium *Escherishia coli* (*E.coli*), free living N fixers—*Azotobacter*, N fixers in legume nodules—*Rhizobium*, and the nitrifying bacteria Nitrobacter and Nitrosomonas. Firmicutes are gram positive, aerobic or facultatively aerobic. They are typically rod-shaped (e.g., *Bacillus*, which also forms endospores), or filamentous like the Actinomycetes, *Streptomyces*.

Bacteria are typically classified by shape (round, rod, spiral), size, whether they are filamentous or not, colony characteristics, whether they are photosynthetic or not, by Gram stain—positive or negative, whether they are aerobic or anaerobic, produce endospores or not, have flagella, and what they can grow on (reflecting metabolism and nutrition requirements). They are now being classified using DNA and RNA analyses.

8.1.2 Fungi

The vast majority of fungi are obligate saprophytes or decomposers; they are aerobic. Some form beneficial symbiotic associations with plant roots known as mycorrhizas, some are endophytic, while others are parasitic. Fungi are eukaryotic, spore-producing, achlorophyllous organisms with absorptive nutrition and they reproduce sexually and/or asexually. The vegetative state of fungi mostly consists of branched mycelium. However, some fungi such as yeasts seldom form mycelium and consist of single cells or collections of cells. Individual branches or filaments of fungi are called hyphae with diameters ranging from 0.5 µm to more than 100 µm. The length of mycelium may be only micrometers in some fungi (e.g., those growing on rock surfaces) to several meters in fungi that produce mycelial strands or rhizomorphs. Growth of the hyphae occurs at the tips of the hyphae. Hyphae of fungi can be made up of many cells partitioned by cross walls (septa) containing one or more nuclei per cell, or they have no septa.

Fungi produce spores that are used for dispersal and in some cases survival; e.g., chlamydospores or thick-walled asexual resting spores. Most spores are airborne [139]. Spores can be sexual or asexual, and can be produced externally or in sacs.

Based on DNA, true fungi include the phyla Glomeromycota, Blastocladiomycota, Chytridiomycota, Zygomycota, Ascomycota, and Basidiomycota plus other fossil phyla. Glomeromycota are arbuscular mycorrhizal (AM) fungi.

Attributes of fungal phyla are summarized below (Note: all fungi are aerobic):

– **Zygomycota**—hyphae without cross walls (septa); asexual spores—sporangiospores, chlamydospores, sexual spores—zygospores. Spores are mostly windborne. Example: bread molds.
– **Ascomycota**—hyphae with cross walls, asexual spores—conidiospores (conidia), arthrospores, yeasts (budding), sclerotia, chlamydospores; sexual spores—ascospores. Spores are mostly windborne. Example: mold fungi, truffles (ectomycorrhizal fungi)
– **Basidiomycota**—hyphae with cross walls and clamp connections; asexual spores—rare (condia); sexual spores—basidiospores. Spores are mostly airborne. Examples: cellulose and lignin (wood) decomposers, mushrooms (ectomycorrhizal fungi)
– **Glomeromycota**—hyphae without cross walls, asexual spores-large single, sexual spores—none. Example: Arbuscular mycorrhizas
– **Chytridiomycota** are mostly saprophytes; many are aquatic. Example: They can cause diseases of agricultural plants and amphibians.
– **Blastocladiomycota** are parasitic on plants and animals, but some are saprobic. They are aquatic and terrestrial and flagellated.

Some microbes with hyphae were originally thought to be fungi, but now are considered to be "fungal-like" organisms—they are aerobic, but tolerate low oxygen. There are two main Phyla in this group:

- **Oomycota**—hyphae without cross walls; asexual spores-zoospores with flagella, chlamydospores; sexual spores—oospores; spores mostly soil/water borne; root pathogens.
- **Mycetozoa** (slime molds).

8.1.3 Soil Algae and Cyanobacteria (Blue-green Algae)

Algae occur in nearly all terrestrial environments on earth and are invariably encountered both on and beneath soil surfaces. The algal flora of the soil includes members of the Cyanochloronta, Chlorophycophyta, Euglenophycophyta, Chrysophycophyta, and Rhodophycophyta. Thirty-eight genera of prokaryotic and 147 genera of eukaryotic algae include terrestrial species, the majority of which are edaphic. Whereas systematic nomenclature of blue-green algae adheres to traditional classification based upon morphological features, proper taxonomic treatment of eukaryotic soil algae is predicated on standard methods of culture and interpretation of physiological attributes, plant mass characteristics, and morphological properties of axenic clones.

Cyanobacteria (also known as blue-green bacteria, blue-green algae and Cyanophyta) are arguably the most successful group of microorganisms on earth and also are the most genetically diverse group. They are unicellular or filamentous, photosynthetic and found in diverse habitats such as on rocks and in soil, water (salt or fresh), hot springs, and even in the Antarctic. Some are diazotrophs (i.e., N fixing organisms that fix atmospheric nitrogen gas into a more usable form, e.g., NH_3) which are either free living diazotrophs such as *Anabaena* and *Nostoc* or they may be symbiotic diazotrophs such as *Anabaena* and *Nostoc* with the gynmosperm cycads, e.g., *Anabaena* with the aquatic fern *Azolla*.

Cyanobacteria produce oxygen gas as a by-product of their photosynthesis and thus are major contributors to not only the global carbon (accounting for 20%–30% of Earth's photosynthetic productivity) and nitrogen cycles but also the oxygen cycle! Cyanobacteria have been theorized to have been the major contributors to changing the Earth's early reducing environment to an oxidizing one allowing for the proliferation of aerobic terrestrial organisms. With the increase in an oxidizing atmosphere, they no doubt contributed to a faster weathering environment for soils via oxidation of the reduced soil iron (Fe^{2+} to Fe^{3+}) and thus contribute significantly to soil development.

8.2   Methods for Determining Soil Microbial Diversity and Populations—Numbers and Biomass

Microbial species and biodiversity can be determined directly by light microscopy on soil particles or on buried slides or nylon mesh. They can also be determined indirectly by growing organisms in culture using soil plates (where soil particles are spread on the surface of media in petri plates), dilution plates, from root washings, and isolation from roots before and after surface sterilization. Another indirect method involves the use of molecular techniques including polymerase chain reaction (PCR), restriction fragment length polymorphism (RFLP), and sequencing (see Section 8.4.5).

8.2.1   Direct Culture, Microscopy and Image Analysis

Culture techniques including soil plating and dilution series have been used to determine the types of bacteria (both anaerobic and aerobic) and fungi that grow in soil and are able to grow on different culture media. However, these techniques have many limitations, including only being able to detect living organisms that grow on specific culture media, such as 2% malt agar. Thus they are not currently used as much as previously and the emphasis is now placed on indirect techniques of assessing soil microbial activity and diversity (see Section 8.4.5).

Microscopy facilitates the determination of size and shape, as well as the numbers of organisms such as bacteria, fungi, yeasts, protozoa, and diatoms. Fluorescent stains make it possible to identify and measure the biota in the soil matrix. Analysis of digitized images by computer software can eliminate much of the operator error in cell identification and measurement. The high cost of microscopes and associated image analysis equipment and the lengthy time involved in analysis are deterrents to the use of this method.

8.2.2   Microbial Numbers and Microbial Biomass

Microbial biomass is a sensitive indicator of many belowground components and interactions. It is a good indicator of environmental toxicity attributable to pesticides, metals, and other anthropogenic pollutants.

Microbial numbers can be determined by the extinction dilution method or the most probable number (MPN) technique. Numbers of bacteria and fungi can be determined from soil dilution plates or the Most Probable Number technique for specific bacteria such as nitrifying and enteric bacteria. Numbers are typically expressed as No. of bacteria or No. of fungal propagules per g of dry soil. However, soil dilution plates can only detect culturable bacteria and fungi so this method is not as commonly used today because of this limitation.

Methods for determining biomass include: (1) microscopy and (2) the measurement of cell constituents released on fumigation. These methods are discussed in more detail in Paul *et al.* [140]. Cell and spore counts can be converted to biomass using microscopy and bacterial cell densities and fungal lengths, but there is now more emphasis on determining microbial biomass for the whole population. Microbial biomass C can be determined using the chloroform fumigation method where the release of CO_2 after fumigation represents that of the labile C in microbes killed by the fumigation. It is typically determined after a 10-day incubation. The respiration rate of unfumigated control soil can be used to determine the background C mineralized [140].

8.3 Mycorrhizas

8.3.1 Types of Mycorrhizas

Mycorrhizas or mycorrhizae (fungus/root associations) are extremely important in maintaining healthy trees in forests. The correct plural of mycorrhiza is mycorrhizas so that term will be used in this section. A good overview of mycorrhizas is presented in Bundrett [141]. All of the commercially important forest trees have mycorrhizas [139]. However, not all plants have mycorrhizas. Aquatic plants typically do not possess mycorrhizas, and even certain terrestrial plants such as those in the mustard family do not form these associations. Many early successional plants, especially those involved with primary succession on sand dunes or after glacial retreat, are nonmycorrhizal or facultatively mycorrhizal [142]. Thus, there is a gradient in mycorrhizal associations from plants that are nonmycorrhizal to those that are obligately mycorrhizal.

There are a number of distinct types of mycorrhizas: arbuscular, ectomycorrhizas, ectendomycorrhizas, arbutoid, monotropoid, ericoid, and orchid [142]. Their features are presented in Table 8.2 along with the major fungal and host taxa. Ectomycorrhizas (EM) are characterized by (1) a sheath or mantle of fungal tissue that encloses the tips of fine roots and (2) the Hartig net or penetration of fungal hyphae between the epidermal and cortical cells. Hyphae of EM fungi do not penetrate cell walls of the host plant and grow between the epidermal and cortical cells. They are mostly in the Phyla Basidiomycota and Ascomycota.

Tab. 8.2 Characteristics of the different types of mycorrhizas.

Characteristic	Ectomycorrhiza	Arbuscular	Ectendomycorrhiza	Ericoid	Arbutoid	Monotropoid	Orchid
Fungi septate	+	−	+	+	+	+	+
aseptate	−	+	−	−	−	−	−
Hyphae enter cells	−	+	+	+	+	+	+
Fungal sheath (mantle) present	+	−	+ or −	−	+ or −	+	−
Hartig net formed	+	−	+	−	+	+	−
Hyphal coils in cells	−	+		+	+	+	+
Achlorophylly	−	−(+)*	−	−	−	+	+
Fungal phyla	Basidiomycota Ascomycota (Glomeromycota)	Glomeromycota	Basidiomycota Ascomycota	Ascomycota (Basidiomycota)	Basidiomycota	Basidiomycota	Basidiomycota
Host taxon	Gymnosperm Angiosperm	Bryophyte Pteridophyte Gymnosperm Angiosperm	Gymnosperm Angiosperm	Ericales Bryophyte	Ericales	Monotropoideae	Orchidales

* Parentheses indicate rare.

Adapted from Smith and Read [142]; Reprinted by permission of Waveland Press, Inc. from *Forest Health and Protection*, Second Edition. © 2011. All rights reserved.

In contrast, hyphae of arbuscular mycorrhizas (AM) in the phylum Glomeromycota do penetrate into cortical cells, but do not penetrate the cell membrane.

Arbuscular mycorrhizas are thus named because they produce arbuscules or branched hyphae-like trees. Some produce vesicles that probably serve as nutrient storage organs. As well as having mycorrhizal associations, some plants have tripartite associations, such as N-fixing plants (e.g., legumes and actinorrhizal plants like red alder). Most of the commercially important genera of trees in the temperate zones are ectomycorrhizal—for example, those in the families Pinaceae, Betulaceae, Fagaceae, Dipterocarpaceae, and Myrtaceae [142]. Ectomycorrhizas usually are visible with the naked eye. Many are dichotomously branched; others are coralloid, or pinnate [142]. Some fungi have loose mantles while others have compact mantles and mantle morphotypes are used to classify mycorrhizas. A number of mycorrhizal fungi also produce hyphal strands or rhizomorphs. However, since many fungi produce similar mantles, and as many as 2,000 species of fungi are thought to be mycorrhizal with one tree species like Douglas-fir, it has been difficult to determine which fungi are actually associated with tree roots in the field. There is usually not a strong association between fungal fruiting bodies and the fungi associated with fine roots, but recent use of molecular techniques has allowed better exploration of this interesting topic.

Arbuscular mycorrhizes are the most abundant mycorrhizes and occur in practically every taxonomic group of plants, including important agricultural plants [142]. Those gymnosperms that do not form EM probably form AM, such as trees in the Cupressaceae (e.g., western red cedar), and in the Araucariaceae and Podocarpaceae in the southern hemisphere. Some species in genera like *Acacia, Casuarina,* and *Populus* have both AM and EM. Interestingly, Douglas-fir is known to have AM in young seedlings. There are far fewer genera of fungi that form AM than EM; the dominant genera are *Glomus, Endogone, Gigaspora, Acaulospora, Entrophospora,* and *Scutellospora.* All of the AM are in the Glomeromycota.

A few fungi form ectendomycorrhizal relationships including reports that they have been found on Douglas-fir [142]. Ectendomycorrhizas may or may not have a mantle, but do possess a Hartig net. The other types of mycorrhizas in Table 8.2 (ericoid, arbutoid, monotropoid, and orchid) have restricted hosts, but many of these hosts, especially in the Ericales, are important understory species in temperate and boreal forests.

8.3.2 Sampling Mycorrhizas

The basic techniques for sampling mycorrhizas are discussed here. Most are adapted from Johnson *et al.* [143]. For further detailed mycorrhizal research see Schenck [144]; Norris *et al.* [145, 146]; Brundrett *et al.* [147, 148, 149]. Goodman *et al.* [150] is a useful publication to learn about identifying EM. An important resource for identification of AM fungi is the International Culture Collection of (Vesicular) Arbuscular Mycorrhizal Fungi [151]. Beg [152] also provides electronic taxonomic aids or computerized "expert systems" for AM fungi.

8.3.2.1   *Sampling Design*

In any research project the first step is to develop a sampling design that assesses the variability and range in populations of mycorrhizal fungi and plants in the ecosystem. This can be done during a preliminary sampling protocol. There is no standard sampling design for mycorrhizas, but if feasible, use individual plants of a particular species as the experimental unit (instead of a plot) located with a stratified-random method. Combining several small samples into one composite sample will reduce within-treatment variability (e.g., collecting and mixing four or five cores at a standard distance from the stem of a plant). Use as many treatment replicates as feasible and employ a factorial experimental design.

8.3.2.2   *Collection of Root and Soil Samples*

Factors that should be considered in developing a sampling protocol are: (1) depth of sampling since mycorrhizal activity varies greatly with depth of forest floor and soil, (2) seasonality, (3) soil properties (e.g., pH, organic matter content, texture, and nutrient content), and (4) plant data (e.g., total cover, shading, species composition, and phenology).

Core extractors or soil probes can be used to collect both root and soil samples. Generally, AM samples require a smaller soil sample volume than EM studies. Use a coarse sieve to collect root fragments in the soil cores. To determine the identity of a host root, trace it back to the plant. Soil and root samples can be stored in Ziploc freezer bags and kept in a cooler or refrigerator until they are processed.

8.3.2.3   *Storage of Samples*

If one needs to know the viability of hyphae, spores, or sporocarps then process the samples immediately after collection. However, if this is not needed, then samples can be refrigerated for several weeks or frozen for long-term storage.

When using trap cultures the timing of sampling during the growing season should be taken into account because the spores of some temperate AM taxa need cold treatment before they will germinate (e.g., 3–8 weeks at 4–5°C). In contrast, isolates of tropical AM fungi, can be extremely sensitive to cold temperatures, and may not survive storage at even 4°C.

8.3.2.4 Determining Mycorrhizas in Samples

Field samples can assessed for mycorrhizas using direct and indirect approaches. It is usually good to use both methods. Using direct methods AM can be quantified microscopically in cleared roots that have been selectively stained to show the AM fungal structures inside the roots [147, 148, 153]. The wet-sieving method can be used to quantify the spores of AM fungi after they are extracted from soil [147]. Flotation-adhesion can also be used [154].

Direct methods can also be used for EM since EM can usually be easily seen by examining root tips with a dissecting microscope. However, some types of EM in *Populus* and *Salix* can be difficult to see. When this occurs, root staining methods used for AM colonization and examination with a compound microscope may be necessary. Ectomycorrhizal fungal sporocarps and mycorrhizal mats (e.g., those formed by *Hysterangium* sp., and *Tricholoma* sp.) also can be directly observed [155]. EM fungi can be identified based on the morphology of root tips, sporocarps, and mycelial mats [147, 150, 156, 157] as well as by using DNA.

The indirect approach involves bait-plant bioassays to detect all viable propagules of mycorrhizal fungi, (e.g., spores, mycorrhizal roots, and extraradical hyphae). This approach better reflects the total activity of mycorrhizal fungi than direct counts of sporocarps, spores, or colonized root lengths [147]. Bait plants are typically grown in diluted field soils and mycorrhizal colonization is determined in the plants after a short time period [158]. Field soils are mixed with sterilized media (e.g., a sand-soil mixture) using a dilution series so that the densities of propagules can be calculated. However, because mixing field soil with the sterilized media can destroy hyphal networks [159, 160], another approach is to sow seeds or plant seedlings of bait-plants directly into undisturbed soil cores [161].

8.3.3 Determination of Mycorrhizal Fungal Species

It is difficult to determine individual taxa of mycorrhizal fungi because the parts of mycorrhizas (e.g., arbuscules, vesicles and hyphae of AM, and the Hartig net, mantle and mycelium of EM) are hard to distinguish morphologically. However, using morphotyping, EM structures can be identified to genus and the AM spores and EM fungal sporocarps can be differentiated to the species level. Molecular and biochemical methods have now been developed to identify mycorrhizal fungal species.

8.3.3.1 Analysis of Sporocarps and Spores

The easiest method for examining the species composition of AM fungal communities is to collect, count, and identify spores. EM sporocarps can also be analyzed. However, there are limitations to sporocarp and analysis because mycorrhizal fungi vary among taxa. Also within taxa there are seasonal and annual variations. There is also a poor relationship between root colonization and densities of sporocarps or spores. Some fungi rarely sporulate while others produce copious spores or sporocarps. Long sampling times (more than 5 years of observations) are required to detect most of the EM fungal species on a given site.

8.3.3.2 Morphotypes of Ectomycorrhizas

Ectomycorrhizal roots mostly have distinguishable macroscopic features, and the easiest method to assess the species composition of EM fungal communities is to identify EM morphotypes. EM fungi are identified from a series of anatomical and morphological features [150, 156, 157], allowing identification of EM fungi to genus and sometimes species.

8.3.3.3 Trap Cultures for Arbuscular Mycorrhizal Fungi

Trap cultures are useful for determining the diversity of AM fungi [162]. They are usually conducted by mixing field soils or roots with sterile sand in pots, adding the seeds of the host plant, and checking the pots after 3–4 months the pots for sporulation. This process may need to be repeated since the number of sporulating species can double or triple after each culture cycle [163].

8.3.3.4 DNA and Biochemical Techniques

Fungal species not distinguished by EM morphology can be distinguished by their DNA [164]. Mycorrhizal fungal species on a plant root as well as the plant species can be identified from the same DNA extract by separately amplifying plant and fungal tissues using polymerase chain reaction (PCR) primers. Biochemical techniques such as fatty acid profile analysis of fungal lipids and immunofluorescence have also been developed for identification and quantification of AM fungi.

There are a number of approaches for identifying ectomycorrhizal fungi using the Polymerase Chain Reaction (PCR) and restriction length polymerase reaction (RFLP). The method used by Cline *et al.* [164] is outlined below. More information on molecular methods for mycorrhizas found in soils is given in Section 8.4.5.

The initial step is isolation of DNA or RNA from tissues (mycorrhizas, sporocarps, etc.). It uses the method described by Gardes and Bruns [165] with the following modifications: the volume of cetyltrimethylammonium bromide (CTAB) lysis buffer is reduced from 300 µL to 40 µL, liquid N is used instead of a dry ice bath, autoclaved plastic micropestles are used (Sigma, St Louis MO, USA), and the DNA extract is re-suspended in TE buffer (10 mM Tris, 1 mM L^{-1} ethylene diamine tetraacetic acid (EDTA), pH 8.0). Extracted DNA samples were stored at $-40°C$ for future analysis.

Randomly selected root tips from different morphotypes are screened by internal transcribed spacer (ITS). To generate RFLPs, Polymerase Chain Reaction PCR is used to amplify the internal transcribed spacers of the nuclear ribosomal RNA gene, using the fungus-specific primer pair ITS-1F and ITS-4. The final reaction mixture consisted of a 1:1000 dilution of the DNA extract, 200 µm each of dATP, dTTP, dCTP, dGTP, 200 nM of each primer, 3 mM MgCl$_2$, 0.5 mg mL^{-1} of sterile bovine serum albumin, 0.05 units µL^{-1} of *Taq* polymerase (various suppliers), and the standard PCR buffer as supplied by the manufacturer. The volume of the final reaction mixture was 25 µL, overlaid with a single drop of mineral oil (Sigma). Sterile water is used as a negative control. DNA from *Tricholoma saponaceum* can be used as a positive control for each run of PCR. Amplification was performed with a denaturation of 95°C for 35 s, an annealing temperature of 55°C for 55 s, and an increasing extension period of 72°C for 45 s plus 4 s per cycle over 36 cycles.

Amplified products are digested with restriction enzymes *Hpa*II, *Cfo*I, and *Rsa*I, separated on 2% agarose gel stained with ethidium bromide and digitally photographed for band analysis using the GELCOMPAR II software package (Applied Maths, Inc., Austin TX, USA). Gel images are processed to eliminate distortion and bending of the gel, and band sizes are calibrated by comparison with a GELCOMPAR II using UPGMA using the fuzzy dice procedure with a band position tolerance of 1. Results are examined and adjusted by hand as needed. Ambiguous RFLP clusters are further evaluated by sequencing of the ITS rRNA gene for up to 10 samples from each RFLP pattern.

For each EM fungal taxon as defined by distinct RFLP patterns based on cluster analysis, sequences are used to identify closely related known taxa based on sequence homology. To obtain reliable identification, one should sequence the ITS rRNA gene using the primer pair ITS-1F and ITS-4. If needed, this is followed by sequencing of approximately 360 bp of the mitochondrial large subunit rRNA gene using primer pair ML5/ML6 as described by White *et al.* [166] and/or approximately 650 bp of the nuclear large subunit rRNA gene using the primer pair LROR/LR16 [167]. Closely related sequences are identified using the National Center for Biotechnology Information web-based BLAST search engine [168].

PHYLIP version 3.6a3 (J. Felstenstein, University of Washington, Seattle, WA, USA) can be used to generate neighbor-joining, parsimony, and maximum-likelihood trees to examine the phylogenetic placement of unidentified EMF taxa with reference to published sequences obtained through BLAST searches as well as sequences obtained from EMF taxa fruiting at the study sites. Sequence homology greater than or equal to 98% for the ITS region and 99% for the nLSU rRNA gene are considered sufficient to assign tentative species-level designations to unidentified EMF taxa.

8.3.4 Ectomycorrhizal Quantification

Detailed methods of quantification of EM colonization of roots are given in Bundrett [149]. Ectomycorrhizal colonization of roots can be assessed by direct examination. Individual plants or soil cores may serve as the sampling units. Sampling individual plants is beneficial because the host can be identified to species, and other information about the host plant can be collected (e.g., age, height, crown diameter, diameter at breast height (DBH), etc.). When roots from many different plants are highly intermingled, however, individual plants cannot be sampled. Thus, soil cores are often used, but extracting roots from soil and organic matter in cores is tedious if the core contains much organic matter [169]. Then it may be advantageous to homogenize and subsample the core before washing.

Materials needed for procedure

1. Soil coring devices with hammer attachment
2. Elutriator

3. Standard soil sieves with various mesh sizes
4. Forceps
5. Razor blades and plastic petri dishes
6. Small trays
7. Microscope slides and coverslips
8. Compound and dissecting microscopes
9. 0.1 mol L^{-1} sodium pyrophosphate
10. Cotton blue in lactic acid (0.05 g cotton blue with 30 mL lactic acid (85%–90%)).
11. Formalin acetic acid (FAA) (5 mL formalin (i.e., 37% formaldehyde solution), combined with 90 mL 70% ethanol, and 5 mL glacial acetic acid).

Procedure

Take samples of uniform volumes of soil systematically or randomly with a soil coring device. Soil cores are placed in Ziploc plastic bags and kept cool until processed. Cores can be stored at 2°C for several weeks or at −20°C for several months without obvious deterioration of mycorrhizas.

Break up the soil core and soak it in 0.1 mol L^{-1} sodium pyrophosphate for approximately 1 h. The core plus pyrophosphate solution is transferred to an elutriator, approximately 200 mL water is added, and then agitated with compressed air for about 1 h. A slow flow of water is used to wash dissolved organic matter and fine soil particles from the sample. Allow the overflow to run through a 0.5 mm mesh sieve to catch small root fragments and mycorrhizas. Roots and mycorrhizas can be extracted from soils using a bucket, water, and a sieve if an elutriator is not available. Place the soil core in a sieve and use a jet of water to remove as mineral material. Be certain that roots are not being lost. After the initial sieve washing, vigorously stir the soil sample in a bucket of water to sink the mineral material to the bottom. The roots and organic debris will float to the surface, and can be poured off and collected on a sieve. If needed a second volume of water can be added and the process repeated until no further root material is acquired.

The material retained in the elutriator or on the sieve is placed in a shallow tray of water and examined with a dissecting microscope. Forceps are used to separate EM roots and root tips from humic materials. Place them in a dish of distilled water. Extracted roots should be kept in water to prevent drying. If samples have a large amount of humus or organic matter they may need to be examined thoroughly.

Use the grid-line intersect method with a dissecting microscope (see Fig. 8.1) to estimate total root length and numbers of EM root tips.

You should confirm that root tips are EM by using a compound microscope. Slides are prepared by squashing individual tips between a coverslip and a microscope slide. A drop of cotton blue in lactic acid is used as a stain. You can also use a hypodermic needle or a razor blade to cut tangential or radial sections that include a small amount of root cortex and mount in cotton blue in lactic acid. Slides should be examined at 400 X to observe the mantle and Hartig net.

1. Measure root length of stained (A) or unstained (B) roots with a dissecting microscope using the gridline intersect method

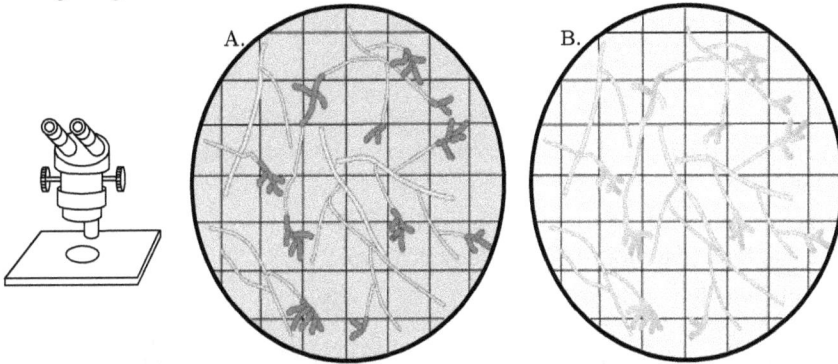

2. Count the length of mycorrhizal roots or the number of mycorrhizal root tips

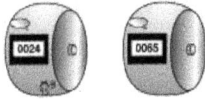

Results: 65 = Root length (cm)
24 = Mycorrhizal root length (cm)
36 = Mycorrhizal root tips

Fig. 8.1 Using the gridline intersection method for estimating total root length and numbers of ectomycorrhizal root tips [148, 149].

Below is the procedure for estimating ectomycorrhizal (EM) fungal diversity based on EM morphotypes.

1. Use dissecting and compound microscopes to sort ectomycorrhizas with different morphological characters. Place samples into small-petri plates with water. Clean ectomycorrhizas can be stored in water in the refrigerator for about a week before they start to decompose.

2. Characterize each morphotype based on external features such as color, texture, rhizomorph structure, and hyphal diameter, and internal features such as mantle thickness, mantle pattern, Hartig net development, specialized cells, hyphal junctions, and emanating hyphae. Refer to Agerer [156, 157], and Goodman *et al.* [150] for detailed descriptions of how to characterize ectomycorrhizas.

3. If the sampling and sorting process is standardized, dried morphotypes can be quantified using a good balance.

4. Photograph samples of morphotypes.

5. Samples of EM root tips can be frozen or lyophilized for DNA analysis or preserved in FAA for future morphological comparisons.

8.3.5   Identification of Ectomycorrhizal Sporocarps

Sporocarp identification provides supplementary data to the identification of EM morphotypes. However, sporocarps are ephemeral, so surveys are limited to the main fruiting seasons.

Materials

1. A measured length of chain attached to a stake or dowel to establish circular plots
2. Digging and cutting tool (e.g., a pocket knife)
3. Truffle fork (e.g., four-tine garden cultivator)
4. Aluminum foil
5. Labels
6. Mushroom collection container (e.g., basket, or plastic bucket)
7. A 3–5 w/v aqueous solution of KOH
8. Melzer's solution. Combine 2 g p-dimethylaminobenzaldehyde (PDAB), 76 mL 95% ethanol, and 24 mL concentrated HCl.

Procedure

1. Select the sampling location and sampling design using multiple forest stands A site should be sampled for a minimum of 5 years and every 1–2 weeks during peak fruiting times. There are two basic sampling designs can be used: (1) convenience sampling and (2) quantitative sampling: If one only wants a list of species, or a rough estimate of total richness in a given habitat, then use convenience sampling involving repeated site visits during the fruiting seasons. Comparisons among sites can be obtained by standardizing the area examined for sporocarps. If variation in species richness among sites is needed use quantitative sampling. Standardize the sample area (use approximately 1,000 m^2 per stand). Rectangular plots divided into contiguous subplots are easiest to relocate if long-term observations are conducted. More species may be detected, however, using small (e.g., 4 m^2) plots dispersed along transects. If hypogeous (belowground species) are to be sampled some plots must be raked with a truffle fork [170]. Disturbed plots may not be able to be used for future sampling, however. A combination of permanent plots and temporary transects can be used when plots are raked.

2. When collecting sporocarps, detach the entire base from the substrate.
3. Collect specimens in all developmental stages, give them a label and then fieldwrap them in aluminum foil, and placed them in a basket, or plastic bucket.
4. Document the morphological features of sporocarps soon after collection. Some taxa can be stored in a refrigerator for several days, while others must be examined on the collection day.
5. Group sporocarps by types, photograph them and identified, them using macrofungi field guides.
6. Dry sporocarps at 40–50°C for two days and stored them. Dried specimens can be used for molecular identification of mycorrhizal fungi.

8.3.6 Quantification of Arbuscular Mycorrhizal Colonization

8.3.6.1 Staining

A staining technique used by Johnson et al. [143] and Bundrett [149] is recommended. The stain, Trypan blue, works well on most roots; however, chlorazol black E or acid fuchsin may be better for some roots. Black India ink can be used as a stain. A dissecting microscope or a compound microscope.

Materials

1. Biopsy cassettes and large beakers to hold cassettes
2. Water bath and heating element
3. 2.5% or 5% KOH.
4. 1% HCl.
5. Alkaline H_2O_2. Use 3 mL 20% NH_4OH (household ammonia) to 30 mL 3% H_2O_2.
6. Trypan blue stain.
7. Plastic petri dishes (8.5 cm diameter) with grid lines etched in the bottom every 13 cm starting 6.5 cm from the edge of the dish.
8. Dissecting and compound microscope.
9. Microscope slides and coverslips.
10. Polyvinyl alcohol (PVA) mountant, channel counter.

Procedure

1. Collect a sample of roots from a known volume and mass of soil.
2. Wash roots and cut them into 2.5 cm pieces.

3. Place a randomly selected subsample of roots of known mass (0.25 g and 0.5 g, fresh weight) in the biopsy cassettes. You may need to place very fine roots between two layers of plastic-coated window screen inside the cassettes.

4. Place cassettes in a beaker containing 2.5% or 5% KOH and cover with aluminum foil. Inside a fume hood warm the KOH in a water bath (70–90°C) for 1 hour to 4 days. This clears the cytoplasm from the cortical cells. Use 5% KOH to clear old, thick, lignified roots. Clear young or fine roots with 2.5% KOH for a shorter time. It may be necessary to replace the KOH solution with fresh solution midway through clearing of very thick or pigmented roots.

5. After roots are cleared or translucent, decant KOH and gently rinse cassettes with tap water. You should change the water at least six times.

6. Roots may still be pigmented. If so, soak them at room temperature for 10–45 minutes in alkaline H_2O_2, and then rinse the cassettes thoroughly with water. Should the roots become darker in H_2O_2, repeat the KOH, H_2O_2 clearing again.

7. Soak roots for 1 hour in 1% HCl to acidifies the roots so that the Trypan blue stain binds well to the AM fungal structures.

8. Decant HCl, add Trypan blue stain to the beaker of cassettes, and warm in a water bath (90°C) for 1 hour.

9. Decant and save stain (usable at least four times) and rinse cassettes thoroughly with water. The stained roots can be stored in a 1:1:1 solution of lactic acid, glycerin, and water. They can also be stored in distilled water in a refrigerator for up to 2 weeks or in a sealed container (dry) in the freezer for long periods of time.

10. Colonization of stained roots can be measured by the grid-line intersection method with either a dissecting or a compound microscope as explained below. A dissecting microscope provides a direct estimate of AM root length. A compound microscope must be used to quantify the abundance of arbuscules, vesicles, coils, and hyphae.

8.3.6.2 Grid-line Intersection Method with a Dissecting Microscope [1]

1. Evenly spread stained roots (of a known mass) across the bottom of an 8.5 cm diameter gridded petri dish containing about 10 mL of water.

2. Using a dissecting microscope at approximately 40 X magnification, follow the lines with your eyes and evaluate each intersection between a line and a root for the presence or absence of AM fungal structures. The mycorrhizal

---

1 This method is illustrated in Fig. 8.2.

status of the root only at the exact point of intersection between the root and the line.

3. Use a channel counter, or simply two hand-held counters, to count the total intersections and the AM intersections.

4. Percent AM root length is then calculated as shown in Equation (8.1):

$$\text{AM root length} = 100 \times [(\text{number of AM intersections})$$
$$/(\text{number of total intersections})] \qquad (8.1)$$

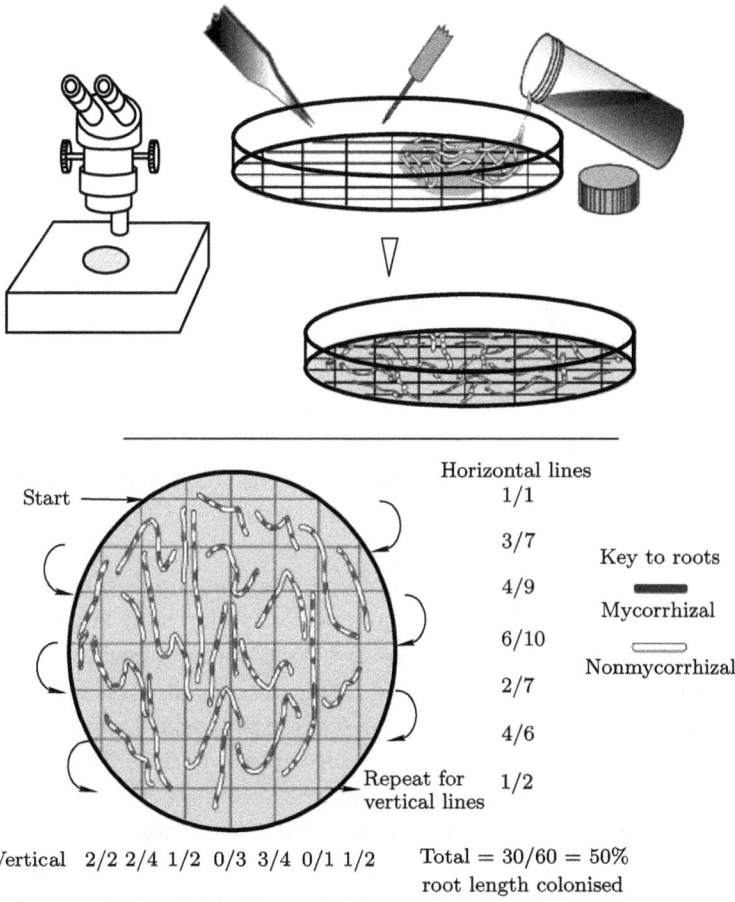

Fig. 8.2 A gridline intersection example using an 8.5 cm diameter round Petri dish with a 1/2 inch (1.27 cm) grid and a 1 m test sample of thread cut into fragments and randomly re-distributed 10 times. Row and column totals are summarized at the bottom of the figure [148, 149].

5. The dimensions of the grid in the petri dish are designed so that if all the lines are examined (both vertical and horizontal), the total number of intersections.

8.4 Indirect Indices for Soil Biological Activity

Indirect indices of soil biological activity include, soil respiration (CO_2 evolution) (see Section 8.4.1), litter decomposition rates (see Section 8.4.2), enzyme activity (see Section 8.4.3), functional biodiversity (substrate utilization profiles) (see Section 8.4.4), and molecular tools (see Section 8.4.5).

8.4.1 Soil Respiration

The techniques discussed here are mainly drawn from Holland *et al.* [171] with modifications. A number of techniques can be used for measuring CO_2 evolution over spatial scales ranging from a single point to square kilometers. At very fine scales measurements can be made using stainless steel or Teflon tubes placed at different soil depths. For scales from 0.1 to 1.0 m^2, place enclosures (or chambers) on the soil surface and measure CO_2 accumulation over time. Micrometeorological measurements on towers are to measure CO_2 fluxes over the largest spatial scales.

Enclosure techniques are used commonly because they are simple, relatively inexpensive, and involve easily moved equipment allowing many locations to be sampled within an ecosystem. There are two basic types of enclosure designs, static and flow-through. Typically static designs use small ports for sampling and a small vent that allows equilibration of outside and inside atmospheric pressures. Flow-through designs are steady-state or non-steady state. In steady state enclosures air is drawn from a known concentration source and carried across the enclosure. In non-steady-state enclosures the CO_2 concentration gradient is reduced because of continual changes in concentration inside the enclosure. Non-steady state static chambers are commonly used and are inexpensive.

Fluxes of CO_2 can be measured by techniques ranging from soda lime and base trap (NaOH and KOH) absorption of CO_2 to use an InfraRed Gas Analyzer (IRGA) system. The soda lime absorption methods tend to underestimate high CO_2 fluxes and overestimate low CO_2 fluxes as a result of varying absorption efficiencies. Soda lime is recommended for accurate 24-hour integrated fluxes.

For enclosure measurement of CO_2 fluxes, IRGA analysis is the fastest. A soil respiration chamber can be linked to an IRGA system [172] (e.g., Licor) when used in the field. A gas chromatograph (GC) can also be used. Two field methods are described here: CO_2 gas sampling and the soda lime technique.

8.4.1.1   CO₂ Gas Sampling

Materials

1. Stainless steel, aluminum or polyvinyl chloride (PVC) permanent collars.
2. Knife for collar location placement.
3. Vented enclosures with a sampling port; Ten mL Nylon or polypropylene syringes with one-way stopcock valves for transporting gas samples to laboratory; The least expensive storage containers are Vacutainer blood sampling vials.
4. Infrared gas analyzer (IRGA) or gas chromatograph (GC).
5. Calibration standards for GC and a canister that can be transported to the fieldsite.

Procedure

1. Permanent collars inserted 5–10 cm into the soil should be installed at least 1 week prior to sampling. Record the time when the enclosure is placed over the soil. Measure the height of the enclosure over the soil surface in at least four places.
2. Take at least ten 7–8 mL air samples with a 10 mL syringe equipped with a one-way stopcock to establish the initial concentration.
3. Then sample 7–8 mL with a 10 mL syringe every 10 minutes for 50 minutes. When each sample is taken record the time. When fluxes are very low sample less frequently over a longer period. With very high fluxes, take samples more often over a shorter period. Be sure that the size of the sample greater than the minimum needed for analysis. Sample the field standard periodically.
4. Analyze the sample using gas chromatography in the lab by injecting it into a GC through a septum.

Flux calculations

Rates of CO_2 exchange are caculated using the difference in CO_2 the concentrations over time. The calculation steps below are from Holland et al. [171]. Measured concentrations are converted to mass units and corrected to field conditions through application of the Ideal Gas Law as shown in Equation (8.2):

$$C_m = (C_v \times M \times P)/(R \times T) \tag{8.2}$$

where

C_m = mass/volume concentration, e.g., µg CO_2–C L^{-1} enclosure, which is equivalent to mg CO_2–C m^{-3} enclosure;

C_v = volume/volume concentration (trace gas concentration expressed as a parts per million or billion by volume, i.e., ppmv or ppbv, respectively, also called mixing ratio), e.g., µL CO_2 L^{-1} enclosure or ppmv CO_2;

M = the molecular weight, e.g., 12 µg CO_2–C μmol^{-1} CO_2;

P = barometric pressure (P is expressed here in atmospheres), e.g., 1 atm; (or 101.325 kPa)

T = air temperature within the enclosure at the time of sampling, expressed as °K(°K=°C+273.15);

R = the universal gas constant (0.0820575 L atm K mole).

The converted concentration values are then used to calculate the flux of interest. The most commonly used equation (Eq. (8.3)) assumes a constant flux (F) and a linear increase in trace gas concentration (C) over time (t):

$$F = V \times C_{rate}/A \tag{8.3}$$

where

F = gas flux as mg CO_2–C m^{-2} h^{-1};

V = the internal volume of the enclosure, including collar volume, expressed as m^3

A = the soil area the enclosure covers, expressed as m^2;

C_{rate} = change in concentration of gas (C_m) over the enclosure period, expressed as mg CO_2–C m^{-3} h^{-1}.

The recommended units for expression of the flux are mg CO_2–C m^{-2} h^{-1}.

A calculation is given below for a 1 hour exposure using a 40 cm diameter × 20 cm tall enclosure (0.025 m^3)

$$C_m = (C_v \times M \times P)/(R \times T)$$

where

C_m = mass/volume concentration (mg CO_2–C m^{-3} enclosure);

C_v, time 0 = 360 ppmv;

C_v, 1 hour = 450 ppmv;

M = 12 µg CO_2–C μmol^{-1} CO_2;

$P = 1$ atmosphere;
$R = 0.0820575$ L atm K mole;
$T = 20°C$.

C_m at time $0 = (360 \times 12 \times 1)/(0.0820575 \times (20.0 + 273.15))$
$\qquad\qquad = 179.58$ mg CO_2–C m^{-3}
C_m at 1 h $= (450 \times 12 \times 1)/(0.0820575 \times (20.0 + 273.15))$
$\qquad\qquad = 224.48$ mg CO_2–C m^{-3}

Below is an example of a flux calculation:

$$F = V \times (C_{rate}/A)$$

where

V = enclosure volume = 0.025 m^3;
$C_{rate} = 224.58 - 179.58 = 44.90$ mg CO_2–C m^{-3} enclosure (change in CO_2
$\qquad$ concentration in 1 h);
A = area of enclosure = 0.126 m^2;
F = flux = $0.025 \times (44.90/0.126) = 8.98$ mg CO_2–C m^{-2} h^{-1}.

8.4.1.2   The Soda Lime Technique

Note: Soda lime gains weight when exposed to CO_2.

The main components of **soda lime** are:
1. Calcium hydroxide—Ca(OH)$_2$ (about 75%).
2. Water—H$_2$O (about 20%).
3. Sodium hydroxide—NaOH (about 3%).
4. Potassium hydroxide—KOH (about 1%).

The method is based on the adsorption of CO_2 by soda lime that is measured by a weight gain. The following absorption reactions occur:

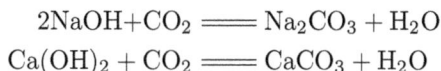

$$2NaOH + CO_2 \rightleftharpoons Na_2CO_3 + H_2O$$
$$Ca(OH)_2 + CO_2 \rightleftharpoons CaCO_3 + H_2O$$

Materials and procedure

1. Obtain soda lime.
2. Dry the soda lime in a clean beaker at 105°C in a drying oven to remove adsorbed moisture.

3. When dry (probably overnight or until it stops losing weight), weigh out approximately 10 g into a soil can (record to at least the nearest 0.001 g).
4. An inverted plastic container or coffee can (e.g., 15 cm diam.) is used as a chamber to trap CO_2 evolving from the soil.
5. At field sites place the soil can with soda lime on the soil and then place the plastic container upside down over it and push its edges into the soil to form a seal around the beaker to trap CO_2 from the soil respiration.
6. Also place a control (blank) sample of soda lime in a soil can in the field also under a plastic container, but one that has a bottom on it (aluminum foil) so that it does not allow CO_2 evolving from the soil to be adsorbed. This control (blank) is treated as all other samples except that it is not exposed to soil CO_2 evolution.
7. Incubate for 24 hours (leave *in situ* so that CO_2 evolution has been subjected to abiotic/biotic fluctuations occurring over the diurnal period).
8. After 24 hours remove the soda lime from under the can and put the top on the soil can to keep CO_2 exchanges from occurring.
9. Dry the soil can of soda lime (uncovered) in the drying oven at 105°C (overnight is sufficient) and then reweigh.
10. Three replicate samples are used at each site as well as one blank.

Calculations

The difference in weights before and after incubation is an estimate of the grams of carbon dioxide evolved from the soil. Multiply this weight by a correction factor of 1.69 (due to 1 mole of water generated by each mole of CO_2 absorbed by the lime) [173]. The units are g CO_2 per container area per 24 hours. This can be converted to g CO_2 m^{-2} h^{-1} as shown in Equation (8.4):

$$S = (W_{sl} \times 1.69)/(A_c \times T) \tag{8.4}$$

where

$S = CO_2$ evolution (g CO_2 m^{-2} h^{-1});
W_{sl} = the soda lime weight gain;
1.69 = correction factor for the calculation of the CO_2 absorption rate of soda-lime;
A_c = the chamber area (m^2);
T = the sampling time in hours.

Do the same calculation for the control (blank) and subtract that value from the sample calculation to derive the correct CO_2 evolution from the soil.

An example calculation is shown below for a 15-cm diameter chamber in Table 8.3.

Tab. 8.3 Example calculation for CO_2 evolution from a 15-cm diameter cylinder in the soil.

	Initial soda lime dry wt. (g)	24 h soda lime dry wt. (g)	Increase in soda lime dry wt. (g)	CO_2 evolution per 24hrs ((soil soda lime dry wt. g – blank g) × 1.69)/(area m^2/24 h)	CO_2 evol. (g CO_2 m^2 h^{-1})
Blank	10.000	10.016	0.116	$((0.278 - 0.116) \times 1.69)$	
Soil	10.000	10.278	0.278	$/(0.0177 \times 24)$	0.644

Notes on correction factors: 1 mole of CO_2 (44 g) reacts with 74 g of Ca[(OH)$_2$] to form 100 g CaCO$_3$ and 18 g H$_2$O in a balanced CaCO$_3$ reaction. The measured increase in soda-lime mass after oven-drying is 26 g (i.e., 100–74). Thus a correction factor of 44/26 (i.e., 1.69) should be applied to the measured mass difference to calculate the true mass of CO_2 absorbed. i.e., for every mole of CO_2 that chemically reacts with soda lime, one mole of water is formed that is subsequently evaporated during drying. Thus, the increase in dry mass after exposure underestimates CO_2 adsorbed by a factor of 18/44 = 40.9%. The correction factor used to account for water formed during chemical absorption of CO_2 by soda lime and released during drying is 1.69 as recognized by Grogan [173] (not 1.41 as originally stated by Edwards [174])

8.4.2 Decomposition Rates of Litter

A number of methods exist to determine litter decomposition rates, including substrate weight loss and CO_2 evolution. Here the focus is on substrate weight loss. Methods vary depending on substrate type. The substrats covered here are fine litter, fine woody debris, coarse wood debris, and standard substrate decomposition. Many of the methods discussed here are drawn from Harmon *et al.* [175] who provide considerably more detail.

8.4.2.1 Fine Litter

Fine litter decomposition rates can be determined by: (1) the litterbag method, (2) litter baskets, (3) tethers to connect litter material, and (4) calculating input-output balances where decomposition rate constants are calculated from the ratio of annual litterfall mass to existing litter layer mass on the ground [176].

The litterbag method is the most common method and is used for leaves, needles twigs, cones, small bark fragments, and fine roots. Mesh size can be vary depending on the litter material.

Materials

The materials needed to construct, place, and retrieve litterbags include:
1. Air-dried litter.
2. Litterbags and monel staples to seal litterbag.
3. Aluminum tags.

4. Flagging.

5. Shovel used for placing below ground litterbags.

6. Nylon line for litterbags tethering.

7. Ziploc bags in which to place litterbags.

8. Paper bags.

9. 50–55°C range drying oven.

10. Hand held GPS unit.

Procedure

1. Construct litterbags (with dimensions of at least 20 cm × 20 cm) of non-degradable materials. To allow access by macro- and megafauna the mesh size openings must be at least 2 mm in size. However, smaller mesh sizes are commonly used depending on the litter material and the placement environment. One mm nylon mesh is often used in low-light, low UV environments and can last several decades. In environments with high levels of UV radiation, use 1.5 mm fiberglass mesh. For very small needles, use 0.4 mm mesh woven polypropylene pool cover or shade cloth. Litterbags can have a larger mesh on the top than the bottom to prevent the loss of small fragments and allow invertebrate entry. Small mesh Dacron sailcloth (50 µm) can be used in low UV environments, especially below ground environments. High UV environments can use woven polypropylene.

2. Place a tarpaulin under trees to collect leaves or needles during the time of peak litterfall. Green material can also be used. Excavate fine roots or obtain them from ingrowth experiments or nursery seedlings.

3. Air-dry litter for at least 1 week and place at least 10 g in each bag. Oven dry sub samples at 55°C to determine moisture content and for chemical analyses. Sew the bag shut with nylon thread or use Monel staples.

4. Record losses during transportation. Each bag should be identified with numbered aluminum tags.

5. Litterbag sampling times vary depending on the ecosystem study objectives, and resources available. In cool and cold environments they should be collected every 3 months (seasonally) for the first year, and every 2 months in warmer environments. Sampling intervals are longer after that (once per year in cool and cold ecosystems). In warm ecosystems, collection should occur every 3–6 months. If possible conduct studies for at least 5 years, i.e., well beyond the usual 1–2 years, to obtain data on the longer transition period from litter to stable soil organic matter.

6. At least four to five litterbags per litter type need to be collected at each sampling time. Avoid pseudo-replication by using single plots [177]. Fewer sampling times and replicates may have to be used to reduce the total number of litterbags involved.

7. Place litterbags on the litter layer surface and pin them down. Insert fine root bags at a 45° angle into the soil profile. Tether bags to lines and flag them both ends for easy retrieval. Draw a sketch map and/or GPS the location of litterbags. After retrieval clean litterbags of adhering particles and place them in Ziploc bags. Samples can be refrigerated for up to 1 week before processing, but if it is more than a week, freeze them.

8. Clean off the litterbag surface, cut the bag open and contents onto a clean sheet of paper or into a large tray. When decomposition is extensive scrape the litterbag insides with a spatula to remove adhering organic matter particles. Remove living plant parts (e.g., roots or moss) as well as rocks and large soil particles. Record the fresh weight of the material and then place the sample in a paper bag. Dry it at 55°C until the mass is stable and determine the dry weight. Take subsamples of the dried material to determine ash free dry weight (450°C for 4 h) and chemical composition. Grind samples to pass a No. 40 mesh sieve. At a minimum determine C, N, and lignin concentration and calculate C:N ratios, and ligno-cellulose index.

8.4.2.2 Woody Debris

Harmon and Sexton [178] thoroughly reviewed the methods for determining decomposition rates of woody debris. The two most frequently used approaches are (1) chronosequences, and (2) time-series. With the chronosequences, age the woody debris pieces in various states of decay and determine density changes through time. Dates can be determined from scars on adjacent trees, living stumps, seedlings and disturbance records (e.g., windstorms, fires, insect outbreaks, etc.).

Chronosequences produce quick data, but there temporal resolution problems because of substitution of space for time. Time series avoid this, but require more time and effort. Woody pieces can be whole logs or shorter sections (e.g., 1 m). Whole log decomposition rates can be determined cutting discs along the length of the woody debris and determining changes in density at each sampling time. Decomposition rates can be determined over very low periods this way, even centuries.

8.4.2.3 Fine Woody Debris

Fine woody detritus includes downed dead branches and coarse roots (> 1 cm). Material is usually < 10 cm diameter and < 1 m long. Samples are not confined like fine litter. Relative to the soil surface they can be suspended, on the soil surface, or buried.

Materials

1. Branches or coarse roots.
2. Chainsaw, hand saw, and clippers.
3. Calipers.
4. Measure or ruler.
5. Aluminum tags and UV resistant ties.
6. Ziploc bags and paper bags.
7. Portable scale.
8. Nylon tethering cord.
9. Hand held GPS unit.

Procedure

1. Use a minimum of 5 sampling intervals over the expected life span of the pieces.
2. Determine diameter, length, radius of the major tissue types (i.e., outer bark, inner bark, sapwood, and heartwood), total volume, and bark cover [179].
3. Use a range of diameters.
4. Collect material from recently fallen trees after windstorms, or fell trees using a chainsaw. Coarse roots can be collected from stumps associated with road or house construction.
5. The initial mass of small pieces can be determined by weighing the entire piece. The diameter at the midpoint and the total length should be measured for each piece. Determine the initial moisture content for each piece using subsamples dried at 55°C until the weight is stable. Trim off the ends that have been exposed to drying for more than a few hours using a chainsaws. Use numbered aluminum tags ties for identification. Determine the initial chemical composition and ash content of subsamples. The same chemical variables used for fine litter should be determined (C, N, and lignin concentration and calculate C:N ratios, and ligno-cellulose index).
6. Piece length should 10 times longer than the mean diameter. Apply an end sealer (e.g., paraffin or epoxy).
7. Avoid pseudo-replication in placement of samples and use appropriate replication. Draw a sketch map or and/or GPS the locations of samples.
8. Collect samples and clean them in the field initially. Finish cleaning in the lab and oven dry them at 55°C until constant weight. Pieces exceeding 5 cm in diameter should be cut into smaller pieces to speed drying. After determining oven-dry weight, grind the material to pass a 40-mesh (0.417 mm) screen and store ground samples for chemical analysis.

8.4.2.4 Coarse Woody Debris (CWD)

Coarse woody debris includes standing and downed boles, stumps, very large branches and large coarse size roots. Decomposition rates of CWD (typically 10 cm diameter and > 1 m long) can be determined by obtaining the volume of the entire piece, calculating fragmentation losses and then removing disks to determine changes in density [175]. Short log sections (< 1 m) can be placed in a wood drying kiln and dried to a constant weight. This may take weeks to dry, however. Decomposition rates can then be determined from mass loss.

Materials

1. Chainsaw, hatchet, hammer and chisel.
2. Tree boles.
3. Calipers (0–150 mm range), diameter tape and tape measure or ruler.
4. Aluminum tags and nails.
5. Ziploc bags and paper bags.
6. Portable scale.
7. Hand held GPS unit.

Procedure

1. Use a minimum of 5 sampling periods.
2. Use the same physical and chemical descriptors of substrate quality used for fine woody debris. The depth and decay type (brown and white rot) and depth of pith should be determined. Estimate total bark cover of bark if it was removed during transport or felling.
3. Use CWD from recently fallen trees after windstorms, or fell trees using a chainsaw.
4. To determine initial mass place short lengths in a large drying kiln. Remove disks from the ends of the CWD to determine density and initial total volume. For initial volume trim 5 cm off the ends if they have dried out and measure the end and middle diameters and total length of each piece. Multiply initial volume and initial density to determine initial mass.
5. Wood and bark are the minimum layers sampled on each piece of CWD. However, separating sapwood from heartwood is recommended [179], since heartwood decays slower than sapwood. Use a hammer and chisel to separate layers.
6. If pieces exceed 2 m long and 25 cm diameter and need to be moved use logging machinery. Draw a sketch map showing the location of the pieces and/or GPS the location of each piece.

7. To avoid pseudo-replication at least three sites should be sampled at each time [177].

8. At each sampling time place short log sections in a wood drying kiln and dry to a constant weight and determine mass loss. For longer pieces determine the remaining bark cover, volume, and density. Estimate total bark cover using a frame of known size and determine the area missing or remaining. The end and middle diameters and total length should be measured to determine the total sample volume. Use a caliper or diameter tape measure maximum and minimum diameters. Correct for fragmentation losses. Using a chainsaw remove a minimum of three systematically spaced disks. Map out and subsample zones appearing different by cutting the disk into pieces [180, 181], and removing subsections. Use subsamples from each piece to determine their density, moisture, and nutrient content.

Use Equation (8.5) to determine the volume of each cylindrical layer.

$$V = \pi r^2 L \tag{8.5}$$

where

V = volume of the cylinder;
r = radius;
L = length of the layer.

The volume of a layer occurring inside the log may be deducted to help calculate the other layers (Fig. 8.3).

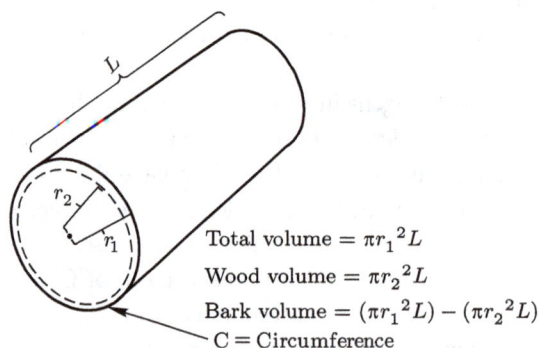

Total volume $= \pi r_1{}^2 L$
Wood volume $= \pi r_2{}^2 L$
Bark volume $= (\pi r_1{}^2 L) - (\pi r_2{}^2 L)$
$C =$ Circumference

Fig. 8.3 Diagram showing examples of calculating different volumes of a log such as total, wood and bark volumes.

If fragmentation of a layer has occurred, then record the radial thickness (r_t) and circumferential length of the layer (C) as well as its longitudinal length (L).

The volume of this layer can be computed as in Equation (8.6):

$$V = r_t CL \qquad (8.6)$$

where
  V = volume of the fragmented cylinder;
  r_t = radial thickness of the fragmented layer;
  C = circumferential length;
  L = longitudinal length of the layer.

The volume of the remaining fragmented layers is calculated like those for unfragmented disks.

It is difficult to remove intact discs from extremely decomposed pieces of logs and which often are elliptical. It is best to cut the disk free and record the maximum radius (r_{max}) and minimum radius (r_{min}) of the disk from the parts that were not removed. The length of elliptical pieces (L) is determined from the removed pieces of the cross section. The volume of the elliptical section is determined using Equation (8.7):

$$V = \pi r_{max} r_{min} L \qquad (8.7)$$

where
  V = volume of the elliptical decomposed log;
  r_{max} = maximum radial thickness of the log;
  r_{min} = minimum radial thickness of the log;
  L = longitudinal length of the elliptical log.

Methods used for layer separation, determination of mass, and subsampling for moisture and nutrient contents are the same as those used for unfragmented disks. Cut subsamples into smaller pieces, place them in paper bags and oven dried at 55°C until mass is constant. Compute the oven-dry weight of the total disk by multiplying the ratio of oven-dry weight to fresh weight of the subsample by the total fresh weight of the layer. Pool subsamples for a given layer and store them.

8.4.2.5   Standard Substrates

Typically standard substrates are used to determine environmental effects on decomposition rates. Common standard substrates are cellulose filter paper, cotton cloth strips, popsicle sticks, and wooden dowels. Cellulose filter paper and hardwood (birch) dowels are good to use because they are very uniform.

Needles and leaves with varying N and lignin concentrations have also been used as standard substrates [182].

Materials

1. Standard substrate (cellulose filter paper or dowels).
2. Nylon sleeves for encasing below ground portion of dowels.
3. Litterbags and nylon thread or Monel staples.
4. Aluminum or UV resistant tags and nylon line for tethering.
5. Flagging.
6. Trowel or shovel.
7. Steel rebar (45-cm long, 6 mm diameter, 45 cm long, and hammer).
8. Ziploc bags.
9. Paper bags to dry litter.
10. Drying oven.
11. Hand held GPS unit.

Procedure

1. Place 5–10 g of cellulose filter paper in a litter bag at several depths in the soil profile. The procedures for recovery and treatment of the decomposed cellulose filter papers follow the procedures for fine litter. Determine the N concentration of the filter paper since it can be quite variable.
2. Place each dowel 30 cm aboveground and 30 cm below ground with a sleeve over the belowground portion. Determine the oven dry mass of each piece. Attach UV resistant flagging or aluminum tag to each portion. Use steel rebar to form a pilot hole when placing the dowel. Mark the area with flagging and draw a location map and/or GPS the location of the site.
3. In temperate forests filter paper can be sampled weekly or monthly. They last less than a year in cool and cold environments and a shorter time in warm tropical environments. Dowels can be sampled yearly in temperate and boreal forests and a more frequently in warm environments [183].
4. Retrieve and process dowels as follows. Find the aboveground portion since it may no longer be attached to the belowground part, and place it in a Ziploc bag. To find the belowground portion, use a trowel or shovel and place it in the same Ziploc bag. Record the lengths of the above- and below ground parts, and note any insect activity. In the lab, remove any soil with a moist paper towel and cut the above- and belowground portions into short (2–5 cm) sections. Dry at 55°C until the mass is constant (5–7 days). Wiley mill and place in storage containers until they can be analyzed.

8.4.2.6   Calculation of Decomposition Rates

Because materials in litterbags are often contaminated with soil litter dry weights have to be corrected before determining mass loss. Ash a subsample for 4 h at 450°C. Mass remaining should be expressed on a percent ash-free dry mass (AFDM) basis of the initial and final litter samples.   Coarse roots and very decayed material also need to be corrected for ash content. Standard substrate data should also be reported in ash-free values. Wood ash contents are low, so there is no need for correction.

There are several models used for fitting the ash free mass loss data [184]. The simplest is the single negative exponential model [176]:

$$X/X_o = e^{-kt} \qquad (8.8)$$

where
X/X_o = proportion of litter mass remaining at time t;
t = time elapsed, expressed as years or days;
e = the base of the natural logarithms;
k = the decomposition rate constant. k values indicate the rate of mass loss; higher k values indicate faster mass loss. This model produces a single decomposition rate constant (k value). However, this model does not accurately describe long-term litter decomposition because it is based on initial rapid decomposition data.

Using least-squares linear regression the negative exponential model can be fitted to the data. For decomposition rate constants, time should be expressed as a fraction of 1 year (e.g., 182 days = 0.5 year). For fast decomposing litter daily or weekly decomposition rates may be used. Least-squares regression gives values for slope (k), intercept (predicted mass remaining at $t = 0$), and coefficients of determination (r^2). Where the single negative exponential model does not fit the data, a double exponential model may be better [184]. This assumes that litter can be partitioned into two types, labile and recalcitrant. This model better fits decomposition of litter types that have a rapid initial mass loss followed by slower mass loss. An third model is the single negative exponential with asymptote, where mass loss declines to zero and a fixed proportion of recalcitrant litter remains (e.g., 30%). This is generally not realistic over long time scales, but it is better for determining the amount of "stable" organic matter produced as a product of litter decomposition.

An example of calculations for determining single exponential k values for Pacific silver fir needle decomposition in the Cascade Mountains of Washington state, USA follows. Figure 8.4 shows mass remaining of needles in litterbags over a 6-year period. Table 8.4 shows mass loss and single exponential k values calculated using the Olsen [176] equation and best fit regression [182] at each sample time.

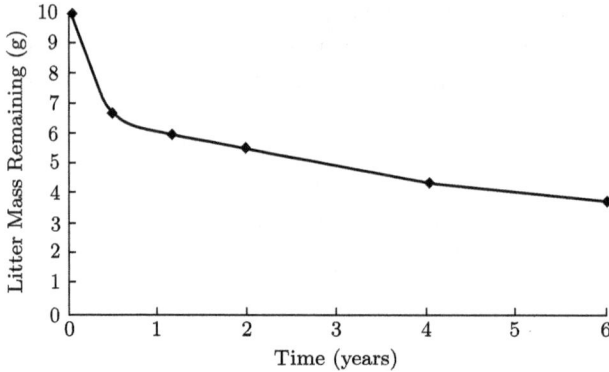

Fig. 8.4 Average ash-free dry mass of Pacific silver fir needles remaining in litter bags at five sampling times in the Cascade Mountains of Washington, USA [185].

Tab. 8.4 Calculations for estimating k (yr^{-1}) for decaying Pacific silver fir needles [185].

Time		Ash-free Dry Mass (g)	k^* (yr^{-1})	k^{**} (yr^{-1})	R^2
Months	Years				
0	0	10.00			
6	0.5	6.66	0.813		
14	1.2	5.88	0.454	0.439	0.86
24	2	5.47	0.302	0.274	0.88
48	4	4.37	0.207	0.173	0.79
72	6	3.65	0.169	0.14	0.84

* Olsen, 1963.
** Best fit regression.

In addition to k values you can also determine from the data set above the:

Time to 50% decay–0.693/k (e.g., 0.693/0.169 = 4.1 years)
Time to 95% decay–2.996/k (e.g., 2.996/0.169 = 17.7 years)
Time to 99% decay–4.605/k (e.g., 4.605/0.169 = 27.2 years)
Mean Residence Time −1/k (e.g., 1/0.169 = 5.9 years)

8.4.3 Soil Enzymes

The methods discussed here generally follow those of Sinsabaugh et al. [186]. More details on soil enzyme analysis can be found in Dick et al. [187]. Determining enzyme activity in soil and litter is relatively simple, but choosing assay conditions and interpreting the results are more difficult. The first decision is whether to measure activities under near ambient conditions or to make standardized comparisons among samples. In the latter case the data represent a measure of the resources that a microbial community is using for decomposition.

Cellulases, hemicellulases, pectinases, phenol oxidases and peroxidases, chitinases, peptidases, and phosphatases are generally the enzymes of interest. Four extracellular soil enzymes are typically studied: acid phosphatase, β-glucosidase, N-acetyl-glucosaminidase, and phenol oxidase. Acid phosphatase (phosphomonoesterase) is responsible for the mineralization of organic N and P to inorganic N (NO_3, NH_4) and P (PO_4) [188, 189]. β-glucosidase (hydrolase) aids in accelerating litter decomposition by breaking down labile cellulose and other carbohydrate polymers. High levels of both acid phosphatase and β-glucosidase indicate high organic matter quality, rapid mineralization, and high nutrient availability. N-acetyl-glucosaminidase (chitinase) degrades chitin, and releases low molecular weight C and N rich compounds. Phenol oxidase (oxidative) is one of several enzymes involved in the degradation of lignin in plant litters and is in higher abundance where litter quality is low or recalcitrant. Organisms most effective at lignin degradation are adapted to low nutrient conditions.

For enzyme analysis 3–5 g (wet weight) of soil from each sample core (< 2 mm fraction) can be sent to a lab for analysis. Activity of β-glucosidase, acid phosphatase, N-acetyl-glucosaminidase, and phenol oxidase are usually determined using adopted from Saiva-Cork *et al.* [190]. Briefly, assays are conducted using 0.5 g soil in sodium acetate buffer (50 mmol L^{-1}, pH 5.0) using either methylumbelliferyl enzyme substrates (for β-glucosidase, acid phosphatase, and N-acetyl-glucosaminidase) or L-3, 4-dihydroxyphenylalanine (L-DOPA, 10 mmol L^{-1}, [Sigma-Aldrich, St. Louis, Missouri, USA]) as the substrate (for phenol oxidase). Four analytical replicates per sample should be used. Assays are conducted at 20°C for 1–2 h, using both substrate and sample controls. Enzyme activity is expressed as nmol h^{-1} g^{-1} of dry weight mineral soil for each sample.

8.4.4   Functional Biodiversity—Phospholipid Ester-linked Fatty Acids (PLFA) and Substrate Utilization Profiles

Functional biodiversity can be studied using biochemical methods. These methods involve the entire community and provide a quantitative description of the structure and function of soil microbes within a particular environment. Two methods are commonly used: (1) characterization of microbial biomass composition by identifying extracted phospholipid ester-linked fatty acids (PLFA) and fatty acid methyl esters, and (2) use of substrate utilization profiles of soil communities to assess community structure. Only Substrate Utilization Profiles developed by Biolog, Inc., [191] are discussed here. More details on PLFA sampling can be found in Sinsabaugh *et al.* [186].

The easy to use Biolog Microbial ID System can rapidly identify over 2,500 species of aerobic and anaerobic bacteria and fungi. The system uses 95 different C substrates plus one negative control in a 96-well microtiter plate. The more complex the microbial community, the more substrates are utilized. A complex of low concentration growth factors are also placed in each well along with a redox dye (tetrazolium violet) that is reduced to purple formazan during microbial activity. This indicates that the C substrate in the well has been utilized. This change can be detected using a microtiter optical density plate reader at 590 nm wavelength. Microplates are incubated at 25°C for up to 120 h and optical density readings are taken are at 24 h intervals.

Biolog, however, only provides information about microbial community members that can be cultured under the provided conditions. Not all microbes may be represented. Furthermore it represents potential metabolic activity and in situ activity. Sometimes the data are hard to interpret.

8.4.5 Molecular Tools for Ecological Systems

Nucleic acid or "molecular" analyses provide the most comprehensive information on microbial community composition. It is the only approach available for estimating the absolute diversity of soil microbiota and in many cases is the only available approach for monitoring the dynamics of specific taxa. Molecular ecology is developing rapidly and new techniques and equipment are being introduced continually. Kits are now available for removing DNA from soil and also for carrying out PCR. Amplified DNA can then be sent to a lab for sequencing.

The following methods are used:
- DNA extraction from soil or plant roots
- Polymerase chain reaction (PCR) which make copies of a gene of interest
- Restriction fragment length polymorphism (RFLP)
- Selection of primers for a particular "target" gene or genes of interest e.g. rDNA, RPB2 (RNA polymerase II), β-tubulin, EF-1α (elongation factor), etc.
- Gel Electrophoresis for viewing DNA strands by length in "bands"
- Sequencing to find the order of nucleotide bases (ATGC) for the gene

8.4.5.1   DNA Extraction from Soil

DNA can be extracted from soil using bead beating, sonication, and enzymatic lysis. Yeates *et al.* [192] provided the following technique for extracting DNA from soil using bead beating:

1. Extraction buffer (100 mL of 100 mmol L^{-1} Tris-HCl [pH 8.0], 100 mmol L^{-1} sodium EDTA [pH 8.0], 1.5 mol L^{-1} NaCl) was mixed with 100 g (wet weight) of soil.
2. Glass beads (100g, Bio-Spec Products, Bartesville, U.S.) were added and the sample blended in a Bead-Beater (Bio-Spec Products) for 2 min.
3. Sodium dodecyl sulphate (SDS) was added (10 mL; 20 %) and blending continued for a further 5 s.
4. The sample was incubated at 65°C for 1 h, transferred to centrifuge bottles (250 mL) and centrifuged at 6,000 g for 10 min.
5. The supernatant was collected, and the soil pellet re-extracted with further extraction buffer (100 mL), incubation at 65°C for 10 minutes and centrifugation as above.
6. Supernatants were transferred to centrifuge tubes (50 mL) containing a half-volume of polyethylene glycol (30%)/sodium chloride (1.6 mol L^{-1}), and incubated at room temperature for 2 h.
7. Samples were centrifuged (10,000g for 20 min) and the partially purified nucleic acid pellet resuspended in 20 mL of TE (10 mmol L^{-1} Tris-HCl, 1 mmol L^{-1} sodium EDTA, pH 8.0).
8. Potassium acetate (7.5 mol L^{-1}) is added to a final concentration of 0.5 mol L^{-1}. Samples were transferred to ice for 5 min then centrifuged (16,000 g, 30 min) at 4°C to precipitate proteins and polysaccharides.
9. The aqueous phase was extracted with phenol/chloroform and chloroform/ isoamyl alcohol and DNA was precipitated by adding 0.6 volume isopropanol.
10. After 2 h at room temperature, DNA was pelleted by centrifugation (16,000 g for 30 min) and resuspended in TE (1 mL).

The most common techniques now involve purchasing extraction kits that remove organic inhibitors (e.g., humic acids) such as the SoilMaster DNA Extraction Kit from Epicentre Biotech [193]. This kit provides all the reagents necessary to recover PCR-ready DNA from a variety of environmental samples, including soils. The kit utilizes a hot detergent-lysis process combined with a simple chromatography step that removes organic inhibitors, such as humic and fulvic acids. More details on this kit are available on line [193].

Other soil DNA extraction kits are:

–   SurePrep^{TM} Soil DNA Isolation Kit from Fischer Bioreagent [194].
–   PowerSoil, PowerMax and UltraClean soil DNA isolation kits from MoBio Labs Inc [195].
–   NucleoSpin Soil DNA Isolation Kit from Clontec [196].
–   FastDNA Spin Kit for soil from MP Biomedicals [197].

8.4.5.2   PCR

PCR is used to amplify a single or a few copies of a piece of DNA thus generating thousands to millions of copies of a particular DNA sequence. It is used for DNA cloning, sequencing, DNA-based phylogeny (functional analysis of genes), and detection and diagnosis of diseases. The method relies on thermal cycling, i.e., cycles of repeated heating and cooling of the reaction for DNA melting and enzymatic replication of DNA. Primers (short DNA fragments) containing sequences complementary to the target region along with a DNA polymerase are the key to enabling selective and repeated amplification.

Almost all PCR applications employ a heat-stable DNA polymerase, e.g., Taq polymerase, which enzymatically assembles a new DNA strand from DNA building-blocks, the nucleotides, by using single-stranded DNA as a template and DNA oligonucleotides (also called DNA primers), which are required for initiation of DNA synthesis. The vast majority of PCR methods use thermal cycling, i.e., alternately heating and cooling the PCR sample to a defined series of temperature steps. These thermal cycling steps are necessary first to physically separate the two strands in a DNA double helix at a high temperature in a process called DNA melting. At a lower temperature, each strand is then used as the template in DNA synthesis by the DNA polymerase to selectively amplify the target DNA. The selectivity of PCR results from the use of primers that are complementary to the DNA region targeted for amplification under specific thermal cycling conditions.

Most PCR methods typically amplify DNA fragments of between 0.1 and 10 kilo base pairs. A basic PCR set up requires several components and reagents. The PCR is commonly carried out in a reaction volume of 10–200 μL in small reaction tubes (0.2–0.5 mL volumes) in a thermal cycler. The thermal cycler heats and cools the reaction tubes to achieve the temperatures required at each step of the reaction. Thin-walled reaction tubes permit favorable thermal conductivity to allow for rapid thermal equilibration.

Typically, PCR consists of a series of 20–40 repeated temperature changes, called cycles, with each cycle commonly consisting of 2–3 discrete temperature steps, usually three. The cycling is often preceded by a single temperature step (called *hold*) at a high temperature ($> 90°C$), and followed by one hold at the end for final product extension or brief storage. The temperatures used and the length of time they are applied in each cycle depend on a variety of parameters. These include the enzyme used for DNA synthesis, the concentration of divalent ions and dNTPs in the reaction, and the melting temperature (Tm) of the primers. The following steps are followed: Initialization, Extension/elongation step, Final elongation, and Final hold.

To check whether the PCR generated the anticipated DNA fragment (also sometimes referred to as the amplimer or amplicon), agarose gel electrophoresis is employed for size separation of the PCR products. The size(s) of PCR products is determined by comparison with a DNA ladder (a molecular weight marker), which contains DNA fragments of known size, run on the gel alongside the PCR products.

There are more than twenty variations of the basic PCR technique. Quantitative PCR is commonly used. Quantitative PCR methods allow the estimation of the amount of a given sequence present in a sample—a technique often applied to quantitatively determine levels of gene expression. Real-time PCR is an established tool for DNA quantification that measures the accumulation of DNA product after each round of PCR amplification.

Quantitative PCR (qPCR)

qPCR used to measure the quantity of a target sequence (commonly in real-time). It quantitatively measures starting amounts of DNA, cDNA, or RNA. qPCR is commonly used to determine whether a DNA sequence is present in a sample and the number of its copies in the sample. *Quantitative real-time PCR* has a very high degree of precision. QRT-PCR (or QF-PCR) methods use fluorescent dyes, such as Sybr Green, EvaGreen or fluorophore-containing DNA probes, such as TaqMan, to measure the amount of amplified product in real time. It is also sometimes abbreviated to RT-PCR (Real Time PCR) or RQ-PCR. QRT-PCR or RTQ-PCR are more appropriate contractions, since RT-PCR commonly refers to reverse transcription PCR (see below), often used in conjunction with qPCR.

Like DNA extraction there are now PCR kits available. For example, the QIAGEN Multiplex PCR Kit is available in a convenient ready-to-use master mix format. The QIAGEN Multiplex PCR Master Mix includes HotStarTaq DNA Polymerase and a unique PCR buffer containing the novel synthetic Factor MP. Together with optimized salt concentrations, this additive stabilizes specifically bound primers and enables efficient extension of all primers in the reaction without the need for optimization. Q-Solution, a novel additive that enables efficient amplification of "difficult" (e.g., GC-rich) templates, is also supplied.

8.4.5.3 Restriction Length Fragment Polymorphism RFLP

RFLP is a technique that exploits variations in homologous DNA sequences. It refers to a difference between samples of homologous DNA molecules that come from differing locations of restriction enzyme sites, and to a related laboratory technique by which these segments can be illustrated. In RFLP analysis, the DNA sample is broken into pieces (digested) by restriction enzymes and the resulting *restriction fragments* are separated according to their lengths by gel electrophoresis. An example of the use of RFLP is given in Section 8.4.4 for determination of species of mycorrhizal fungi. RFLP is not as commonly used now due to the rise of inexpensive DNA sequencing technologies.

8.4.5.4 Primers

Primers are strands of nucleic acid that serve as starting points for DNA synthesis. The length of primers is usually not more than 30 (usually 18–24) nucleotides and they need to match the beginning and the end of the DNA fragment to be amplified. They direct replication towards each other—the extension of one primer by polymerase then becomes the template for the other, leading to an exponential increase in the target segment.

PCR primers designed using sequences of the internal transcribed spacer (ITS) regions of ribosomal DNA (rDNA) are commonly used in diagnosis of soil bacteria and fungi. The genes of 16s rRNA subunits are used for bacteria and archaea, while the genes of subunits of 18S, 5.8S, and 28S are used for fungi. Other marker genes used for fungi are β-tubulin, EF-1α, and RPB2 genes. Species determination and phylogenetic surveys are carried out using PCR amplification.

8.4.5.5 Gel Electrophoresis

Gel electrophoresis is a method for separation and analysis of DNA and RNA and their fragments, based on their size and charge. It is used to separate a mixed population of DNA and RNA fragments by length, to estimate the size of the fragments. Nucleic acid molecules are separated by applying an electric field to move the negatively charged molecules through an agarose matrix. Shorter molecules move faster and migrate farther than longer ones because shorter molecules migrate more easily through the pores of the gel. DNA Gel electrophoresis is usually performed for analytical purposes, often after amplification of DNA via PCR, but is also used as a preparative technique prior to use of other methods, such as RFLP and DNA sequencing, for further characterization.

If several samples have been loaded into adjacent wells in the gel, they will run parallel in individual lanes. Depending on the number of different molecules, each lane shows separation of the components from the original mixture as one or more distinct bands, one band per component. Incomplete separation of the components can lead to overlapping bands, or to indistinguishable smears representing multiple unresolved components. Bands in different lanes that end up at the same distance from the top contain molecules that passed through the gel with the same speed, which usually means they are approximately the same size. Molecular weight size markers are available that contain a mixture of molecules of known sizes and they are run on one lane in the gel parallel to the unknown samples. The bands observed can be compared to those of the unknown in order to determine their size.

After the electrophoresis is complete, the molecules in the gel can be stained to make them visible. The most common dye used to make DNA or RNA bands visible for agarose gel electrophoresis is ethidium bromide (EtBr). It fluoresces under UV light. By running DNA through an EtBr-treated gel and visualizing it with UV light, any band containing more than ~20 ng DNA becomes distinctly visible. SYBR Green I and SYBR Safe are other dsDNA stains (produced by Invitrogen - www.invitrogen.com). They are more expensive, but 25 times more sensitive.

After electrophoresis the gel is illuminated with an UV lamp in a light box. EtBr flouresces reddish orange in the presence of DNA and can be photographed with a digital camera. The DNA band can also be cut out of the gel, and can be dissolved to retrieve the purified DNA. However, even short exposure to UV light causes significant damage to DNA. Thus if the DNA is to be used after separation on the agarose gel, it is best to avoid exposure to UV light by using a blue light excitation source, e.g., the XcitaBlue UV to blue light conversion screen from Bio-Rad. A blue excitable stain is required, such as one of the SYBR Green or GelGreen stains. Blue light is also better for visualization since it is safer than UV (eye-protection is not such a critical requirement) and passes through transparent plastic and glass. This means that the staining will be brighter even if the excitation light goes through glass or plastic gel platforms.

8.4.5.6   DNA Sequencing

DNA sequencing is the process of determining the nucleotide (AGTC) order of a given DNA fragment. To date, most DNA sequencing has been performed using the chain termination method developed by Frederick Sanger [198]. However, new sequencing technologies such as pyrosequencing are now commonly being used.

8.4.5.7 Metagenomics and Transcriptomics

Pyrosequencing has allowed researchers to quickly and inexpensively identify all of the transcribed genes in a soil sample (the transcriptome) [199]. It allows identification of both culturable and non-culturable microorganisms down to genus and even species levels, by mass sequencing part or all of their collective genomes (metagenome), or even allows identification of what genes are transcribed in a metagenome (metatranscriptomes) [200]. These techniques are rapidly supplementing and even replacing the traditional role of isolating and culturing bacteria and fungi from their natural habitats or observing them microscopically, while providing a more complete picture of microbial populations and their interactions and functions under natural conditions.

8.5 Soil Invertebrates

Soil invertebrates live in mostly aerobic environments and fragment litter, graze on fungal and bacteria and are predators on other soil invertebrates. They are important for processing energy in soil systems and can be used to determine responses of ecosystems to perturbations. The array of species constituting soil invertebrates is very large, encompassing virtually all terrestrial invertebrate phyla. Typically they are discussed in terms of body width (microfauna, mesofauna, and macrofauna) as shown below. Numerous vertebrates (moles, etc.) are also important in soil ecosystem functions, but they are not discussed here.

1. Macrofauna (> 2 mm):
   - Isopoda (isopods, pill bugs, wood lice, potato bugs)—crustaceans—7 pairs of legs
   - Diplopoda (millipedes)—2 pairs of legs per body segment
   - Chilopoda (centipedes)—1 pair of legs per body segment
   - Scorpions
   - Araneae (spiders)
   - Opiliones (harvestmen)
   - Insecta (insects)
   - Isoptera (termites)
   - Hymenoptera (ants)
   - Coleoptera (beetles)

- Annelida (earthworms)
- Gastropoda (slugs and snails)—molluscs

2. Mesofauna (100 μm–2 mm):
- Collembola (springtails—primitive insects—0.1 mm to 2 mm long)
- Acari (mites—0.1 to 2 mm long—8 legs)
- Enchytraeidae (enchytraeids)
- Diplura (diplurans—primitive blind insects with antennae—2–5 mm long)
- Pseudoscorpions (2–8 mm)
- Symphilids—white 11–12 pairs of legs—no eyes)
- Pauropoda (white terrestrial myriopods segmented 9–11 legs)
- Protura (wingless primitive insects in family Apterygota)

3. Microfauna (< 100 μm) (aquatic animals):
- Protozoa (protozoans)
- Rotifera (rotifers)
- Tardigrada (tardigrades)
- Nematoda (nematodes)

Image analysis and access to the Internet have made the identification of microscopic biota easier. Molecular techniques are now commonly used to assist in taxonomic identification.

The three faunal groups—macrofauna, mesofauna and microfauna—are extracted differently. Methods for all soil invertebrates, however, are not covered in this section. Most of the methods discussed here are drawn from Coleman *et al.* [201] with modifications. Macrofauna are often collected by hand or sometimes with mechanical assistance. The focus here is on methods used for sampling litter and soil for earthworms. Handsorting and flotation can also provide data on populations on other macrofauna like molluscs and soil-dwelling invertebrates, like insect larvae. Colonial insects, e.g., termites and ants, require more specialized sampling methods (see Brian [202] and Lee and Wood [203]). Mesofauna are extracted using heat, water-density gradients, or flotation while culture or centrifuge techniques are usually used for extraction of microfauna.

Invertebrate populations are mostly reported on an area basis, i.e., number m^{-2} to a certain depth, or number m^{-3}, rather than number g^{-1} or kg^{-1} of soil. Soil corers or other collection tools like hand trowels should be thoroughly cleaned to prevent cross-contamination of samples. Alcohol or Clorox can be used to sterilize sampling tools for microfauna. For meso- and macro-arthropods, simple cleaning with a brush or paper towels is enough. Presterilizing hand trowels eliminates the time factor needed for cleaning.

8.5.1 Macrofauna with Emphasis on Earthworms

Macrofauna include macroarthropods (isopods, millipedes, centipedes, scorpions, spiders, harvestmen, and insects), earthworms and mollusks. Populations of surface-dwelling macroarthropods are often determined by pitfall trapping. Open-top traps are set into the ground level with the surrounding soil surface. This method is useful for surveys, or assessing the relative activities of macroarthropods, but it is not a good quantitative technique. Hand sorting a known area is standard better method for quantifying surface macroarthropod densities on a per unit area basis. Typically sample areas of 1 m^2 are used. Subsurface macroarthropods, such as insect white grubs and cicada nymphs, are assessed using the same hand-sorting method recommended for earthworm sampling below.

There are three major ecological groupings of earthworms (epigeic, endogeic, and anecic) based on their burrowing and feeding strategies. Epigeic species live in or near the surface litter and feed primarily on coarse particulate organic matter. Endogeic species live within the soil profile and feed primarily on soil and associated organic matter. They mostly have temporary burrow systems that are filled with cast material. Anecic species, such as *Lumbricus terrestris*, live in more or less permanent vertical burrow systems that may extend 2 m into the soil profile. They feed primarily on surface litter which they frequently pull down into their burrows. They may also create "middens" at burrow entrances. Middens consist of a mixture of partially incorporated surface litter, earthworm castings or feces, and soil.

8.5.1.1 Extraction Methods

The two main approaches to sampling earthworm populations are physical and behavioral. With physical methods a known volume of soil is sorted through mechanically or manually. Behavioral methods rely on earthworm response to an irritant applied to the soil, e.g., a solution of dilute formalin. This causes the earthworms to rise to the surface. They can then be manually collected where they can be collected manually. The two methods work differently for different earthworm species. Sometimes it may be advantageous to use both methods since this may provide more accurate data then either method alone. Passive methods are focused on here.

Passive methods of earthworm sampling are the most common and include hand-sorting, washing and sieving and flotation. To obtain a soil sample use a spade or removed or a mechanical device, such as a hydraulic coring tool. It can be difficult to remove replicable volumes of soil using a spade. To obtain a

consistent sample mark the edges of your sampling unit by driving a 25 cm ×
25 cm steel quadrant driven into the ground. Then use a square-point spade to
sample driven vertically into the soil. Use a sampling depth of 20–30 cm.

Separate earthworms from the soil either by hand-sorting or washing and
sieving. In the field place the soil on plastic sheets and carefully sort through
it. The soil can also be returned to the laboratory for sorting. Soil should be
hand-sorted quickly because any worms killed during digging will decay rapidly.
Soil can also be or stored at 4°C until sorting. Although simple hand-sorting is
laborious and small worms < 2 cm in length are easily overlooked. Small worms
contribute significantly to earthworm numbers, but are not a large proportion
of total earthworm biomass. Washing and sieving can supplement hand-sorting
to correct for very small specimens or to collect earthworm cocoons. You can
use a standard soil sieve or more elaborate mechanical washing/sieving devices.
The size of the mesh size should prevent small earthworms or cocoons from
washing through. Soils with high clay contents may need to be soaked in a
0.5% metaphosphate solution to facilitate sieving. To preserve the worms add
formalin to the solution.

Washing and sieving work better than hand-sorting when sampling dense fibrous root systems. Physical sampling methods are best for an accurate. Population assessment of shallow-dwelling earthworm species such as those belonging
to the lumbricid genera *Aporrectodea* and *Octolasion* or the native American
megascolecid genus *Diplocardia*. Species such as *Lumbricus terrestris*, however,
which form permanent burrows as long as 2 m deep, are difficult to sample by
physical methods. Small immature specimens of *L. terrestris* may be sampled
by hand-sorting. Adults, are rarely recovered by digging or coring methods.
Behavioral sampling methods must be used for an adequate assessment of populations of species like *L. terrestris*. This is important since *L. terrestris* can
comprise the majority of earthworm biomass at some sites.

Earthworm data

Earthworms can be identified to species using keys (e.g., Dindal [204]). Population data are usually expressed as total biomass or number of earthworms per
unit area, (e.g., g m^{-2} or numbers m^{-2}). Use ash-free dry mass (AFDM) rather
than oven-dry mass. AFDM corrects for varying mass of soil in the earthworms'
intestines. This can account for 50%–70% of earthworm dry mass. AFDM is
determined by combusting oven-dried worms (60°C to constant weight), or a
ground subsample, in a muffle furnace at 500°C for 4 hours (g AFDM = g dry
wt − g ash wt).

Physical sampling methods may produce data that contains body parts, as
well as whole individuals. If one is interested only in biomass estimates then
this isn't a problem, but it is for determination of numbers of individuals. To

overcome this problem count any fragment of a whole worm that contains the anterior portion (head) as an individual. Do not count body fragments that lack a head as individuals, but include them in total biomass estimates. Changes in land use, glaciation, and exotic earthworm species introductions have affected the distribution of native North American species of earthworms [205]. In many locations some exotic species will be are usually present [206]. Introduced species generally seem have different ecological roles than native species and may respond differently to ecosystem treatments and changing environments (e.g., global warming). In areas where species replacements are recent or ongoing, it is to document long-term dynamics of earthworm populations.

8.5.2 Mesofauna

8.5.2.1 Mites and Collembola

Mites and Collembola are microarthropods and dominate the mesofauna populations; they generally constitute 90%–95% of microarthropod samples. Oribatid mites (Cryptostigmata or Oribatei) are often the most abundant and diverse mites reported from temperate deciduous and coniferous forests. Mites and Collembola are extracted by physical methods or high gradient extraction.

Extraction by Physical Methods

Physical extraction techniques are based on dispersing soil samples in a solvent. Soil invertebrates will float based on density differences (salt or sugar solutions) or on the affinity of the arthropod cuticle for organic solvents. Different solutions and modifications have been proposed, including a heptanes flotation techniques proposed by Geurs et al. [207].

High-gradient extraction

High-gradient extraction depends on the behavioral responses of microarthropods to temperature and moisture gradients to drive them out of soil cores and into collection funnels [208].

Many variations of the high-gradient extraction method have been proposed, some of which require elaborate heating and/or cooling systems. The most common high-gradient extractor is a Tullgren funnel, also known as a Berlese funnel or Berlese trap (Fig. 8.5).

Antonio Berlese described this method of dynamic sampling in 1905 with a hot water jacket as the heat source. In 1918 Albert Tullgren described a

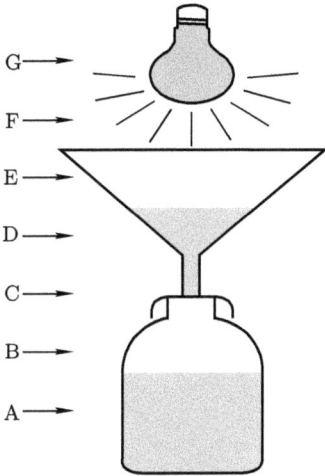

Fig. 8.5 A schematic of the Tullgren or Berlese funnel to extract soil invertebrates from the soil [209].

A = preservative liquid, B = receptacle, C = rubber stop, D = soil or litter, E = funnel, F = heat, G = electric lamp.

modification, where the heating came from above by an electric bulb. The heat gradient was increased by placing an iron sheet drum around the soil sample. The Tullgren funnel works by creating a temperature gradient over the sample such that mobile organisms will move away from the warmer temperatures and fall into a collecting vessel, where they perish and are preserved for examination. Figure 8.5 [209] shows how it works: a funnel (E) contains the soil or litter (D), and a heat source (F) e.g., an electric lamp (G) heats the litter. Animals escaping from the desiccation of the litter descend through a filter into a preservative liquid (A) in a receptacle (B). The soil sample is placed on a mesh sieve that will allow the soil animals to pass but retain most of the soil particles.

A relatively simple and inexpensive extraction system (Fig. 8.6), which is a variation of the Tullgren funnel, is described below. It is portable and can easily be adapted to a variety of situations (cold room, refrigerator, or bench top extraction).

Materials needed for procedure

The basic extractor is constructed of the following:
1. Styrofoam and plastic trays
2. Plastic cups and caps
3. PVC tubes
4. Heat lamps

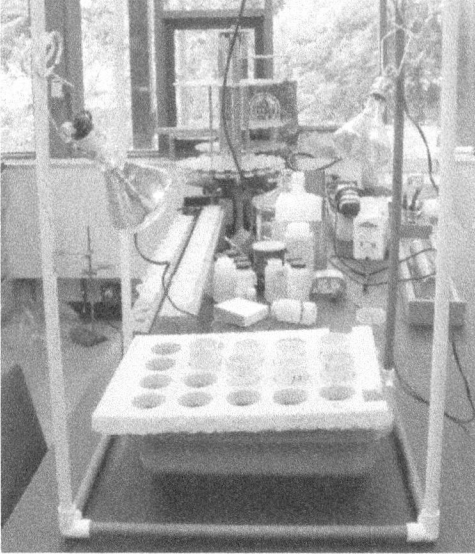

Fig. 8.6 High gradient invertebrate extractor.
Photo by Robert Edmonds.

5. Window screen material (1 mm mesh), which can be attached to the cores with rubber bands
6. Vials
7. Cheesecloth
8. Antifreeze—polyethylene glycol

Procedure process

1. Wrap the window screen material around the base of the PVC tube attach with rubber band and trim excess.
2. Wrap soil core in a double layer of cheese cloth, trim and place in PVC tube.
3. Place 20 mL of antifreeze in plastic cups.
4. Place plastic cups in trays beneath the holes in the Styrofoam sheet.
5. Insert the PVC tubes containing soil through the holes in the Styrofoam sheet and insert them into the plastic cups with antifreeze.
6. Be sure to write the sample ID information in permanent ink on the PVC tube and the plastic cup.
7. Once all the soil cores are in place heat lamps at the appropriate height over the samples in the PVC pipes to obtain a strong heat gradient. The lamps are attached to a PVC support bar.

8. Set the entire unit into a refrigerator or cold room at about 10°C. This is recommended for development of a large temperature gradient, it is possible to extract soil microarthropods at room temperatures if a cold room or refrigerator is not available.
9. An extraction time of 5 days is usually sufficient for most soil types.
10. At the end of the extraction period, carefully disassemble the extractor and remove the plastic cups and cap the.
11. Once capped, the samples can be stored.

As well as Mites and Collembola, other arthropods will be sampled by high-gradient extraction, including pseudoscorpions and protura. The enumeration and identification of extracted microarthropods will vary, depending on the goals of the study and the degree of taxonomic expertise available. Many microarthropods can be identified to suborder or family under a dissecting microscope. Finer taxonomic resolution typically requires mounting individual specimens on microscope slides, or use of an inverted compound microscope.

8.5.2.2 Enchytraeids

Enchytraeids are also extracted from soil samples using exposure to heat and light, but in a wet system rather than dry. They are commonly sampled with 5 cm diameter soil corers, to a depth of 5 cm. Soil is sliced at depths of 2.5 cm increments and placed on a modified wet-funnel extractor. The soil, placed on a sieve in a funnel filled with water, is exposed to increasing heat and light. After 4 h of presoaking (to saturate the soil), the light intensity in 40 watt bulbs is gradually turned up (over approximately 3 h period) on a rheostat timer until the soil surface reaches a temperature of 45°C. Enchytraeids respond by moving away from the heat and light and passing through the sieve into the water below. They are then counted and/or preserved in 70% ethanol. To determine the ash-free soil dry weight, subsamples are freeze-dried and then ashed in a muffle furnace at 500°C for 4 h.

Materials needed for procedure

1. Funnel rack similar to that used for nematode extraction
2. 8 cm diameter plastic funnels attached to collection vials
3. Screening or cheesecloth
4. 40 watt lightbulbs on a single rheostat, positioned 2–4 cm above soil samples

Procedure

1. Place pre-wetted soil sample on screening or cheesecloth inside the funnel.
2. Add water to barely cover the wet soil sample.
3. Over a 3-hour extraction period gradually increase light intensity to warm the surface soil temperature to 45°C.
4. Preserve enchytraeids captured in water beneath the funnels in 70% ethanol.

Send specimens to a taxonomic expert for identification below the family level.

8.5.3 Microfauna

8.5.3.1 *Protozoans*

There are four ecological groups of protozoans: the flagellates, naked amoebae, testacea, and ciliates. Most of them are difficult to observe in soil samples and require culture techniques to quantify their presence, including the use of culture wells or microtiter plates. A most probable number procedure involving presence-absence is often used to assess protozoan numbers.

8.5.3.2 *Rotifers*

The top few centimeters of soil is the best habitat for rotifers. Since they live in soil water films soil samples should not be dried since for identification live specimens are needed. Vegetation samples can be agitated in water and the suspension washed through a sieve, similar sizes used for nematodes. Rotifers are easily extracted using centrifugation techniques. A dissecting microscope at 400 X magnification or higher is used for identifying living rotifers. The majority of identified soil and litter rotifers fall in the order Bdelloidea.

8.5.3.3 *Tardigrades*

Tardigrades are small (100–500 μm in length) and have cylindrical bodies that are flattened on the ventral side. They have four pairs of clawed legs. They include carnivores, as well as bacterial and algal feeders. Some have unknown feeding habits. They occur in soils and vegetation, mostly in wet or moist habitats. To sample remove vegetation, including mosses and lichens gently from soil or rock surfaces and dry in a paper bag until ready for processing. Mineral soils can be sampled with a core sampler. Surface leaves or organic matter can be sampled with a trowel.

8.5.3.4   Microfauna-Nematodes

Soil nematode extraction methods require water. Elutriation, sugar centrifugation, sieving, or misting are commonly used. Extraction should be from fresh soil (or from soil refrigerated less than 2 days). Samples can be processed in the field or laboratory. Nematodes are typically aggregated near their food base (e.g., fungi, roots, protozoa, and insects). Soil type and texture, as well as other properties influence nematode abundance and diversity.

Use the sugar centrifugation and elutriation techniques when 500–1,000 g samples are available. With the centrifugation technique, nematodes are mixed with water and centrifuged to concentrate the nematodes. They are then remixed in a sugar solution of a density at which they will float; this suspension is then centrifuged, and the nematodes are left in a clean solution of sucrose, washed into water. This a fast method and both living and dead nematodes are sampled.

Materials

1. Sucrose solution (454 g sucrose L^{-1} water)
2. Centrifuge tubes (50 mL)
3. Sieves (400-mesh [38 μm], 500-mesh [26 μm], 40-mesh [380 μm])
4. Table centrifuge 1,000 mL plastic beakers
5. 150mL beakers or test tubes, ring stand with funnel
6. 5% formalin solution if samples are to be preserved

Procedure

1. Place 100–200 cm^3 of soil in a large plastic beaker, and weigh it. Remove any rocks with forceps. Add soil to 800 mL of water.
2. Stir for 30 s using a spatula. Pour the solution through 40-mesh on top of a 400-mesh wet sieves and rinse.
3. Remove the top sieve and make sure all the water has filtered through.
4. Backwash the nematodes into a funnel with a 50 mL centrifuge tube beneath, to catch the nematodes.
5. Rinse the funnel with water, centrifuge the liquid for 5 min at 1,750 rpm and decant off all liquid.
6. Fill with the sugar solution.

7. Gently stir the nematode pellet at the bottom of the tube with a spatula until it breaks up and is suspended.
8. Centrifuge for at 1,750 rpm for 1 min.
9. Decant tube contents onto a wet 500-mesh screen.
10. Rinse with water and backwash into a beaker.
11. If you wish to preserve samples use a hot 5% formalin solution added to an equal volume of nematode-water solution.

Identification and calculations of numbers

Trophic groups (i.e., bacterial, fungal, and algal feeders, omnivores, and plant parasites) can be identified based on morphology and known feeding habits using a microscope at 100 X or 400 X. Dilutions can be made if densities are too high. However, dilutions may cause an underestimation of species richness.

Determine the soil moisture and calculate the number of nematodes kg^{-1} of dry soil. If soil bulk density is known, calculate numbers m^{-2} to a 10 cm depth.

8.6   Nitrogen Transformations

Soil microbes are involved in a number of soil N transformations, including N fixation, denitrification, nitrification and N mineralization. Bacteria are mostly responsible for the first three transformations, while bacteria, fungi and soil invertebrates are involved in N mineralization. Transformations of nitrogen can be studied either in the field or laboratory. In this section methods are provided for N fixation and denitrification. Nitrification and mineralization are beyond the scope of this chapter. Only a brief summary is provided here but more details are provided in Robertson *et al.* [210] and Norton and Stark [211].

Nitrification is the microbial conversion/oxidation of NH_4-N to NO_3-N. Common nitrifying bacterial genera are *Nitrosomonas* (which oxidizes NH_4-N to NO_2-N) and *Nitrobacter* (which oxidizes NO_2-N to NO_3-N); several other genera are involved. Fungi can also produce NO_3-N, particularly in forest environments, but the metabolic pathway is different. Ammonium-N in soils generally results from mineralization of N during organic matter decomposition and net N mineralization is the balance between gross N mineralization and microbial N immobilization plus *in vitro* losses of N over an incubation period (e.g., 2 weeks). Some of the organic N mineralized to NH_4-N will be oxidized to NO_3-N by nitrifiers. Net N mineralization is calculated as the change in the sum of NH_4-N plus NO_3-N nitrogen at the end of the incubation time. Nitrification alone can be assayed by considering only the net rate of NO_3-N increase during the incubation period.

Techniques for measuring N mineralization and nitrification include (1) *in situ* incubations of enclosed soils in cores, in which inorganic nitrogen accumulation is measured at the beginning and end of a 2–6 week incubation period (results are expressed as mg N kg^{-1} d^{-1} or basis for g N m^{-2} d^{-1}); (2) laboratory incubations under standard moisture and temperature conditions in which inorganic N accumulation is monitored at 7–30 day intervals for up to a year or more; and (3) isotopic incubations during which changes in a ^{15}N-labeled inorganic N pool is measured over the course of a 1–3 day incubation. Other methods for assessing N availability have been used, including ion exchange resin bags inside intact cores. Replication is very important and generally more than 6 cores are needed. Soil cores are typically thin-walled cylinders (PVC or steel), 24 cm long by 2.5 cm inside diameter. In some ecosystems, e.g., old-growth forests, net nitrification may be zero during an incubation period leading to the conclusion that there is no nitrification in this ecosystem. In this case determining gross nitrification as well as net nitrification will provide better insight into the nitrification process [211].

8.6.1 Nitrogen Fixation

Nitrogen fixation is performed by a wide variety of soil microbes (bacteria, cyanobacteria and archaea) and molecular techniques have revealed more and microbes that are capable of biological N fixation [212]. It occurs in a wide variety of substrates including soils, sediments, rhizospheres, leaves, stems and large woody debris. The bacteria include heterotrophic bacteria, that can be free living or symbiotic, and autotrophic cyanobacteria which can also be free living or autotrophic. Free living photosynthetic bacteria can also fix N. Archaea have recently been discovered to fix N [212].

Many genera of bacteria are involved with N fixation. Free living heterotrophic bacteria can be aerobic (e.g., *Azotobacter* and *Beijerinckia*), facultatively anaerobic (e.g., *Bacillus* and *Klebsiella*), or anaerobic (*Clostridium* and *Desulfvibrio*). Symbiotic autotrophs can be nodule producing (e.g., *Rhizobium* in legumes, *Frankia* in non-legumes like red alder, and *Rhizobium* and *Nostoc* (Cyanobacteria) in other non legumes) or non-nodule producing found in plant rhizospheres (e.g., *Azotobacter* and *Azospirillum*), the phyllosphere (*Azotobacter* and *Klebsiella*) and on woody stems. Free living autotrophic N fixers are mostly cyanobacteria (e.g., Nostoc and Anabaena) or photosynthetic bacteria. *Nostoc* and *Anabaena* also form non-nodule symbiotic associations with lichens, ferns, liverworts mosses and *Nostoc* forms nodule symbioses with cycads, which are gymnosperms.

There are basically three types of methods for determining N_2 fixation rates [213]: (1) acetylene reduction, (2) N accretion through time, using a chronosequence approach, and (3) ^{15}N isotope techniques—isotope dilution, natural abundance and N_2 incorporation [213]. The three methods are outlined below, but details are only given for the acetylene reduction method.

8.6.1.1 Acetylene Reduction

Nitrogenase, the enzyme involved with N_2 gas fixation by microbes, is also involved in reducing acetylene gas to ethylene gas. Measurement of ethylene production is inexpensive, sensitive and easy. The acetylene reduction method works best for short-term assays, but several short-term assays can be integrated to provide a long-term estimate of N_2fixation. It can be used for determining N_2 fixation rates in litter, soil, woody debris, and root nodules, etc. However, a conversion factor must always be determined empirically for a given system using $^{15}N_2$ incorporation. The materials, procedure and calculations generally follow those of Myrold et al. [213].

Materials

1. Acetylene gas
2. Gas chromatograph with a flame ionization detector and Porapak Q column
3. Gas tight incubation containers. e.g., a canning jar sealed with a sealable lid with a rubber septum inserted in it
4. Other materials: Plastic syringes of various sizes for the injection of acetylene and gas sample collection; Gas samples can be stored for a short time (< 1 h) in 1 rnL plastic, but for longer storage use evacuated leak-proof vials

Procedure

1. Place the samples in gas-tight incubation container. Normally 5–10 samples are taken.
2. Add acetylene. Also have control containers (one with acetylene and without a sample and one without acetylene to determine background ethylene production).
3. Take gas samples periodically for analysis of acetylene and ethylene.
4. Determine the volume of the headspace and sample dry weight at the end of the incubation.

Calculations

Ethylene production rates and acetylene reduction rates are calculated from the ethylene concentration versus time data using four steps:

1. Contaminating ethylene contained in the acetylene is corrected for by subtracting the ethylene contaminant concentration (C_c) from the ethylene concentration of each sample incubated in the presence of acetylene (C) to obtain the contaminant-corrected ethylene concentration (C_a) for each sample. Calculation of the contaminant-corrected ethylene concentrations in the acetylene reduction method:

$$C_a = C - C_c \tag{8.9}$$

The rate of contaminant-corrected ethylene production is calculated using linear regression ethylene production in the presence of acetylene to give the background-corrected rate of ethylene production of contaminant—corrected ethylene concentrations against time of sampling. The slope (P_a) is the rate of increase in contaminant-corrected ethylene production.

2. The contaminant-corrected ethylene production rate is adjusted for background production of ethylene in the absence of acetylene. Calculate the slope (P_b) of background ethylene production from the linear regression of background ethylene concentration with time of sampling and then subtract this rate of background ethylene production from the rate of (P). The following equation is used to calculate the ethylene production rate in the acetylene reduction method:

$$P = P_a - P_b \tag{8.10}$$

3. The acetylene reduction activity (ARA) is then calculated on a unit weight (e.g., nmol g^{-1} d^{-1}) or area basis as follows:

$$\text{ARA} = P \times (H/D) \tag{8.11}$$

where
      ARA = acetylene reduction activity;
      P = ethylene production rate;
      H = headspace volume;
      D = sample dry weight or surface area.

The amount of N_2 fixed can be calculated by dividing ARA by the empirical (using $^{15}N_2$) or theoretical ratios of moles of acetylene reduced per mole of N_2 fixed. The theoretical stoichiometric conversion factor is uncertain so simply report ARA when the $^{15}N_2$-derived factor is unavailable.

8.6.1.2 N Accretion through Time

This approach involves determining (1) N accumulation through time (often decades) in ecosystem components, or (2) N accumulation along a chronosequence. It has mostly been applied to unmanaged systems and has been conducted relative to successional processes, e.g., after glacial retreat and during sand dune succession which essentially start with little or no soil N. Accumulation of N is the net difference between losses and gains. In natural ecosystems net accumulation is generally attributed to N_2 fixation with the assumption that other gains and losses balance each other out. This method is relatively insensitive, but it does give information on the magnitude of long-term N fixation. It is usually not applied to managed forests, but it has been used in agricultural systems over a growing season for determining symbiotic N fixation involving nodulated plants, including red alder stands. In this case fixation rates are determined by comparing nodulated plants and nonfixing controls.

8.6.1.3 ^{15}N Based Methods

The three ^{15}N based methods are $^{15}N_2$ incorporation, ^{15}N isotope dilution, and ^{15}N natural abundance. They are described below. For detailed methods consult Myrold *et al.* [213].

^{15}N incorporation

The presence of N_2 fixation can be determined by measuring the incorporation of ^{15}N into a sample exposed to $^{15}N_2$. It provides a short-term rate and needs a closed incubation system. It is generally better for lab than field studies. Also it is not as sensitive as the acetylene reduction method. Researchers need a mass spectrometer, $^{15}N_2$ gas which is available from commercial sources, a closed airtight incubation vessel, a plant growth chamber, plastic syringes and needles for transferring gases, a drying over and a grinder.

N_2 fixation by ^{15}N isotope dilution

The ^{15}N isotope dilution method involves labeling the soil, or nutrient solution, with $^{15}N_2$ instead of atmospheric N_2 and can be used in the field as well as the lab. It provides an integrative measure of N_2 fixation, which is more difficult to do using the $^{15}N_2$ incorporation or the acetylene reduction methods. You need a mass spectrometer, ^{15}N-enriched fertilizer, reference plants, a drying over and a grinder.

^{15}N natural abundance

Plants with symbiotic N_2-fixing bacteria typically have enriched ^{15}N abundance relative to N_2 in the atmosphere. It is intermediate between that of the atmosphere and that of reference plants that only take up N from the soil. Thus one is able to determine the proportions of N from N uptake from the soil and that from N_2 fixation. If the total N content of both types of plant is known then amount of N_2 fixed by bacteria in plant nodules can be determined from the proportions. It is not a very precise measure, however. Vegetation and soils vary spatially and seasonally in ^{15}N abundance. At best it is a semi-quantitative measure of N_2 fixation, when there is 2‰–40‰ difference between the ^{15}N abundance of the soil and the atmosphere. If the difference is greater better quantitative estimates can be determined.

The methods for choosing reference plants and performing the isotope dilution calculations are the same for those of the ^{15}N isotope dilution method.

8.6.2   Denitrification

Denitrification is the reduction of nitrate (NO_3^-), nitrite (NO_2^-) and nitrogen oxides are reduced to the gases nitric oxide (NO), nitrous oxide (N_2O) and dinitrogen (N_2)[214]. Robust quantitative estimations of denitrification rates have been hindered by high spatial and temporal variation in the field. Rates commonly vary over two or three orders of magnitude and are often dominated by very high rates of activity in very small activity centers where oxygen is low and nitrate and carbon availability are high.

There are many available methods for measure denitrification including: (1) acetylene-inhibition, (2) ^{15}N tracers, (3) direct N_2 quantification, (4) N_2:Ar ratio quantification, (5) mass balance approaches, (6) stoichiometric approaches, (7) methods based on stable isotopes, (8) *in situ* gradients with atmospheric environmental tracers, and (9) molecular approaches [214, 215]. The N_2:Ar ratio quantification and stoichiometric approaches are mostly used in aquatic or ocean ecosystems so they will not be discussed here.

Because it is difficult to measure denitrification and because measurements are highly variable the concept of determining denitrification potential was developed. Measuring denitrification potential involves increasing denitrification rates above natural rates under lab conditions by using amendments and wet slurried conditions. This reduces the variability of the process.

The most widely used method for determining denitrification potential is denitrification enzyme activity (DEA). Although DEA is not well related to daily or even hourly denitrification rates it has proven very useful for comparison of treatments, etc.

8.6.2.1   The Acetylene Inhibition Method

The acetylene inhibition method is the most commonly used. Acetylene inhibits the reduction of N_2O to N_2. Thus N_2O is the terminal product of denitrification and can be easily measured. The most critical problems with this method are acetylene inhibition of the production of NO_3 via nitrification which leads to underestimation of denitrification rates, and the failure of the inhibition at low NO_3 concentrations. Lack of diffusion of acetylene to active sites in soil is another problem.

Extracted soil cores are commonly used, but cores create disturbance effects. Thus the chamber method was developed, where chambers are placed over the soil surface. Accumulation of N_2O under the chamber or in circulating can then be measured.

8.6.2.2 ^{15}N Tracer Methods

Several methods based on ^{15}N have been applied in studies of denitrification in soil including isotope fractionation, isotope dilution, ^{15}N mass balances, and direct measurement of ^{15}N labeled gases upon addition of $^{15}NO_3$ and $^{15}NH_4$. However, $^{15}N_2$ flux techniques are expensive and time-consuming to perform.

8.6.2.3   Direct N_2 Quantification

Direct measurements of N_2 emissions from soils are difficult because it not possible to quantify small changes in N_2 concentrations due to denitrification relative to the high atmospheric background of N_2. However, direct flux techniques have application in certain cases, particularly where the use of the acetylene inhibition technique cannot be applied.

8.6.2.4   Mass Balance Approaches

Terrestrial N fluxes can be determined by mass balance which involves tracing the movement of $^{15}NH_4$ or $^{15}NO_3$ into different ecosystem pools. Denitrification is quantified as the ^{15}N not accounted for at the termination of the experiment. It is not very accurate or precise.

8.6.2.5   Stable Isotope Approaches

Stable isotopes involved in denitrification are, for example, those of H and O in water, N and O in NH_4, NO_3, NO_2, N_2O, and dissolved N_2 gas, S and O in SO_4, C in oxidized and reduced dissolved C, O in dissolved O_2 gas, and others. Denitrification causes changes in the concentrations and isotopic compositions of many aqueous species and solids which can be measured.

8.6.2.6   Approaches using in situ Gradients in Environmental Tracers

This method involves determining denitrification rates using analyses of atmospheric environmental tracers as indicators of elapsed time. Common environmental tracers are chlorofluorocarbons (CFCs), sulfur hexafluoride (SF_6) and ^{14}C. This approach is particularly useful for estimating denitrification rates in groundwater.

8.6.2.7   Molecular Approaches

Denitrification involves bacteria, archaea, and fungi and many functional genes. Most studies of bacterial denitrification in natural environments have utilized NO_2 reductase genes [215].

Appendices

Appendix A   Statistics Example

To illustrate the meanings of common statistical terms, a random set of 100 numbers (**observations**) was generated. This set comprises the population, and has the following parameters (as opposed to statistics, which concern samples taken from this population):

- population size, $N = 100$
- population mean, $\mu = 99.582$
- population standard deviation, $\sigma = 9.773$
- population $CV = 9.81\%$

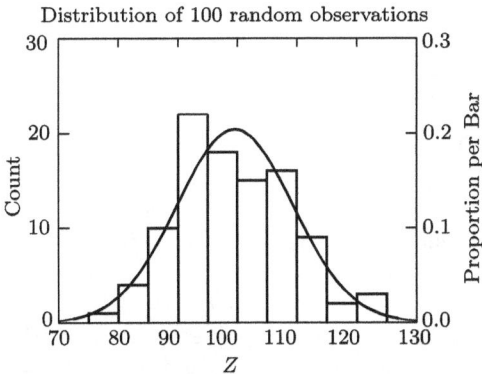

Distribution of 100 random observations

Fig. A.1 Frequency distribution of 100 normally distributed random observations, with mean = 99.582 and standard deviation = 9.773.

Figure A.1 shows a frequency distribution of the population. Note that most of the values fall near the population mean, but a few are relatively far away. To be more precise, 68 values (68%) are within $\pm 1\sigma$ of the mean, 95% are within $\pm 2\sigma$, and 99.7% are within $\pm 3\sigma$, meaning that 5% of the observations are more than 2 standard deviations away from the true population mean. So, if you make one observation on this population (i.e., take a random sample of 1), your estimate of the mean has a 68% probability of being within 9.773 of the true mean (i.e., no more than 9.81% "off"). Pretty good, huh? You might not want to bet the farm on your observation being "right", though, because there is a 5% chance that you could be off by more than $2 \times 9.773 = 19.546$ (i.e., at least 19.63% off)!

The only way to be absolutely certain of μ (disregarding the likelihood of measurement errors) is to measure the entire population. Since this practice is usually out of the question, we typically take a sample of the population, and calculate various statistics (mean, $\bar{x}$; and standard deviation, s) to estimate the population parameters (true mean, μ, and standard deviation, σ). As the sample size, n, gets larger (approaching N, the population size) we reduce (but cannot eliminate) the probability of error (**sampling error**).

Figure A.2 shows the distribution of 20 means of 5 observations each. Note that they are clustered much more tightly around the population mean.

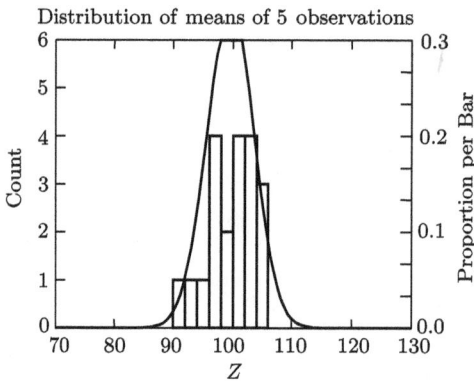

Fig. A.2 Frequency distribution of 20 means of 5 observations each.

These 20 means have a mean of 99.582 (not surprisingly, it's the same as the original population mean), and a standard deviation of 3.918—substantially less than the original population standard deviation of 9.773. In fact, the population standard deviation, 9.773 divided by the square root of the sample size (5) equals 4.371, which is not too far from the observed value of 3.918. You may recognize the term $s_{\bar{x}}$ as the **standard error of the mean**, which is an estimate of the standard deviation of a set of means, with n observations in each mean. In plain(er) English, any mean of 5 observations has a higher probability of being close to the true population mean than does any single observation ($n=1$). Such a mean is a sample from a population of all possible means of 5 observations; this population has a mean of $\mu = 99.582$ (the same as the original population mean), and a standard deviation of the mean equal to $s_x = s/(\sqrt{5}) = 9.773/2.236 = 4.371$, or the original standard deviation divided by the square root of the sample size. In practice, we don't know the population standard deviation. When we take our sample of size n, we estimate it by the sample standard deviation, divide by the square root of n, and use this standard error to calculate confidence intervals around our estimate of the mean.

Similarly, we can take 10 samples of 10 observations each. The 10 means still have a mean of 99.582; the standard deviation is 2.337. The corresponding estimate is 3.090. The two numbers aren't identical, because we only took 10 means of 10, a small sample of all possible means of 10. But you can see that the numbers calculated from this example approximate the theoretical values that the books say you should expect. Figure A.3 shows the distribution of the means of 10.

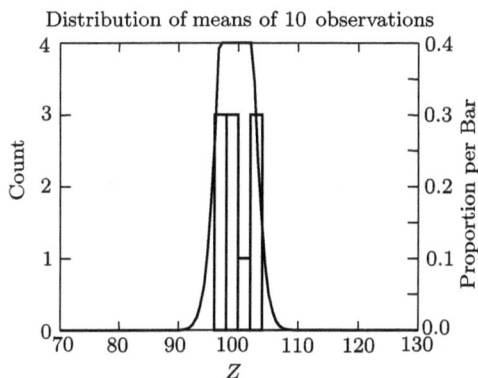

Fig. A.3 Frequency distribution of 10 means of 10 observations each.

Appendix B   Quality Assurance/Quality Control

"Analytical chemistry without QC is guesswork." [216]

Introduction

Quality assurance (QA) is defined as "a system of activities whose purpose is to provide to the producer or user of a product or service the assurance that it meets defined standards of quality with a stated level of confidence" [217].

In the context of a research project involving laboratory chemical analysis, a QA program will give you (the producer or the final user of the data) some confidence that the numbers you have generated are relatively free of errors due to sample contamination, instrument bias, etc., and will help you put limits on the uncertainty of your data.

Taylor further divides quality assurance into two components: quality control and quality assessment. **Quality control** (QC) refers to the routine practices carried out in all phases of the project (field sampling, sample storage and preparation, and laboratory work) to keep data quality "in control," i.e., free (relatively speaking) from errors due to contamination, poor instrument calibration, loss of sample during transport and storage, etc. Such practices include use of distilled-deionized water and reagent-grade (or purer) chemicals, wearing gloves during sample handling, calibrating instruments daily, acid-rinsing glassware, etc. **Quality assessment** covers the tests that you perform to determine that you are, in fact, in control of quality, and includes such things as analyzing method blanks and Standard Reference Materials (SRMs) with each batch of samples, running calibration check samples periodically to verify that instrument calibration is stable, and taking part in inter-laboratory tests. Each sample collection, preparation, and analysis procedure will have its own QA/QC protocol. The use to which the data are going to be put will also determine the extent of the QA/QC steps required.

Appendix C   Example Format for Field and Laboratory Data Sheets

Tab. C.1   Example of a Field Data Recording Sheet.

Field Data Recording Sheet

Carbonates: (*fizz* test) - Use cold dilute ($\sim$2.87 mol L^{-1}) hydrochloric acid (about a 1:10 dilution of concentrated HCl). Effervescence may not be always observable for sandy soils. Dolomite [Ca, Mg(CO$_3$)$_2$] reacts to cold dilute acid slightly or not at all and may be overlooked. It can be detected by heating, by using more concentrated acid, and by grinding the sample. The effervescence of powdered dolomite with cold dilute acid is slow and frothy and the sample must be allowed to react for a few minutes [65]

—— *noneffervescent* (no bubbles form)
—— *very slightly effervescent* (few bubbles form)
—— *slightly effervescent* (numerous bubbles form)
—— *strongly effervescent* (bubbles form low foam)
—— *violently effervescent* (bubbles form thick foam)

(Continued)

---

pH (use field pH kit): _____

Texture (use hand-texture method): _____

Mottling:
    quantity % of area covered (*few* is <2%, *common* is 2%–20%, and *many* are ⩾20%) __
    _____

    size (*fine* is <2mm, *medium* is 2–5mm, and *coarse* is 5–20mm, *very coarse* is 20–76mm,
    and *extremely coarse* is ⩾76mm) _____

    contrast (*faint, distinct, prominent*) _____

Color:
dominant (field moist) _____

        (wet)   _____

        (dry)   _____

mottle (field moist)   _____

        (wet)   _____

        (dry)   _____

Odor: (eg, sulphurous, petrochemical) _____

Animals (earthworms, etc): _____

Structure (or **structureless**): _____

    type (*wedge, platy, prismatic, columnar, angular* or *subangular blocky, granular, single
grain, massive*) _____

    size (*very fine, fine, medium, coarse* and *very coarse*) _____

    grade (*structureless, weak, moderate* and *strong*): _____

Roots:
    type (*clear, suberized, mycorrhizal, nodules*, etc) and **location** (*between peds, in cracks,
throughout, in mat at top of horizon, matted around rock fragments*) _____

    quantity—average across 3–5 representative vertical plane unit areas (*very few* is <0.2/unit
area, *moderately few* is 0.2–1/unit area (or *few* is <1/unit area), *common* is 1–5/unit
area, and *many* are ⩾5/unit area) _____

    diameter size (*very fine* is <1mm, *fine* is 1–2mm, *medium* is 2–5mm, *coarse* is 5–10mm,
and *very coarse* is ⩾10mm) _____

Clay films, bridges, coats, stress surfaces, slickensides, etc: _____

Tab. C.2 Soil Data Summary Sheet.

Soil Data Summary

Soil ID code _____

Sand, % _____

Silt, % _____

Clay, % _____

Textural class _____

Munsell color (dry) _____

Munsell color (wet) _____

Mottling? _____

Available water _____

pH (10:1 in DW) _____

pH (in $CaCl_2$) _____

pH and EC (sat. paste) _____

Exchangeable cations (cmol[+]/kg):

Ca _____

Mg _____

K _____

Na _____

Al _____

ECEC (cmol[+]/kg) _____

Total CEC (cmol[+]/kg) _____

LOI % _____

Total carbon % _____

Total nitrogen % _____

Extractable NH_4^+–N, mg/kg _____

Extractable NO_3^-–N, mg/kg _____

Extractable PO_4^{3-}–P, mg/kg _____

"Total" P, mg/kg (digest) _____

Pyrophosphate Fe, % _____

Pyrophosphate Al, % _____

Citrate-dithionite Fe, % _____

Citrate-dithionite Al, % _____

Tab. C.3 Soil Water Concentration Data Sheet.

Soil Water Concentration Data Sheet

Site _____   Name _____   Date _____

Sample ID	Soil can no. (#)	(T) Tare* wt (g)	(T+S+M) Tare+ Moist Soil (g)	(T+S) Tare+ Dry Soil (g)	(S+M) Moist Soil (g)	(S) Dry Soil (g)	(M) Soil Moisture (g)	Moisture by dry wt (%)

---

* Tare=soil can+can lid

Tab. C.4 Soil Bulk Density Data Sheet.

Soil Bulk Density Data Sheet

Site _____   Name _____   Date _____

Sample	Core type	Core volume (cm³)	Soil can ID	Tare wt (g)	Moist Soil (g)	Rock wt (g)	Dry Soil (g)

References

[1] Hannah L, Carr JL, Landerani A. Human disturbances and natural habitat: A biome level analysis of a global data set. Biodiversity and Conservation 1995; 4: 128-155.

[2] GLP (Global Land Project). Science Plan and Implementation Strategy. IGBP Report No. 53/IHDP Report No. 19. Stockholm, Sweden: IGBP Secretariat, 2005.

[3] Jacobsen T, Adams RM. Salt and silt in ancient Mesopotamian agriculture. Science 1958; 128: 1251-1258.

[4] Hillel DJ. Out of the Earth, Civilization and the Life of the Soil. New York, NY, USA: The Free Press, 1991.

[5] Wild A. Soils and the Environment: An Introduction. Cambridge, UK: Cambridge University Press, 1993.

[6] Lorenzi R. Civilization collapsed after cutting key Trees—Nazca. Discovery Communications, 2009. (Accessed August 11, 2013, at http://news.discovery.com/history/archaeology/nazca-civilization-collapse-trees.htm)

[7] Diamond J. Collapse: How Societies Choose to Fail or Survive. New York, NY, USA: Penguin Group, 2005.

[8] Kennett DJ, Breitenbach SFM, Aquino VV, et al. Development and disintegration of Maya political systems in response to climate change. Science 2012; 338: 88-791.

[9] MEA. Millennium Ecosystem Assessment: Ecosystems and Human Well-Being Synthesis. Findings of the Condition and Trends Working Group of the Millennium Ecosystem Assessment. Washington, DC, USA: Island Press, 2005.

[10] Vogt KA, Patel-Weynand T, Shelton M, et al. Sustainability Unpacked: Food, Energy and Water for Resilient Environments and Societies. UK: Earthscan, 2010.

[11] FAO (Food and Agriculture Organization). World Soil Resources Report 90. Land Resource Potential and Constraints at Regional and Country Levels. Rome, Italy: UN FAO, 2000.

[12] Innes JL. Acid rain, air pollution and forest decline (CASE 5.2). In: Vogt KA, Honea JM, Vogt DJ, et al, eds. Forests and Society: Sustainability and Life Cycles of Forests in Human Landscapes. Wallingford, England, UK: CABI International, 2007: 109-110.

[13] Vogt DJ, Sigurðardóttir R, Zabowski D, Patel-Weynand T. Forests and the carbon cycle (Chapter 6). In: Vogt KA, Honea JM, Vogt DJ, et al, eds. Forests

and Society: Sustainability and Life Cycles of Forests in Human Landscapes. Wallingford, England, UK: CABI International, 2007: 188-218.

[14] Johnson MG. The role of soil management in sequestering soil carbon. In: Lal R, Kimble J, Levine E, Stewart BA, eds. Soils and Global Change. Chelsea, MI, USA: Lewis Publishers, 1995: 351-363.

[15] COMET-Farm Team, 2013. (Accessed August 11, 2013, at http://cometfarm. nrel.colostate.edu/)

[16] Brady NC, Weil RR. The Nature and Properties of Soils. 14th ed. Upper Saddle River, NJ, USA: Prentice Hall, 2007.

[17] WRI (World Resource Institute). Environmental Change and Human Health. Report Series: World Resources 1998-99. Washington DC, USA: World Resources Institute, United Nations Environment Programme, United Nations Development Programme, World Bank, 1998-1999.

[18] Ehui SK, Hertel TW. Testing the impact of deforestation on aggregate agricultural productivity. Agric Ecosystems Environ 1992; 38: 205-218.

[19] Gardner E. Peru battles the golden curse of Madre de Dios—Attempts to reduce the environmental and health impacts of mining cause unrest. Nature, 2012; 486: 306-307.

[20] Gomiero T, Pimentel D, Paoletti MG. Environmental impact of different agricultural management practices: Conventional vs. organic agriculture. Critical Reviews in Plant Sciences 2011; 30: 95-124.

[21] Unpublished data from Vogt KA, Gmur SJ, Vogt DJ, et al. Pan-tropical natural forests assessed from above and belowground: A meta-analysis of soil and climate influences on total net primary productivity. 2013.

[22] Hunter JM. Geophagy in Africa and in the United States: A culture-nutrition hypothesis. Geographica Review April 1973; 170-195.

[23] Lehmann J, Pereira da Silva J Jr, Steiner C, Nehls T, Zech W, Glaser B. Nutrient availability and leaching in an archaeological Anthrosol and a Ferralsol of the Central Amazon basin: Fertilizer, manure and charcoal amendments. Plant and Soil 2003; 249: 343-357.

[24] ScienceDaily. Amazonian Terra Preta Can Transform Poor Soil Into Fertile. Material retrieved from Cornell University March 1, 2006. (Accessed June 22, 2013, at http://www.sciencedaily.com/releases/2006/03/060301090431.htm)

[25] Petersen RG, Calvin LD. Sampling (Chapter 2). In: Klute A, ed. Methods of Soil Analysis. Part 1. Physical and Mineralogical Methods. 2nd ed. Madison, WI, USA: Soil Science Society of America, Inc, 1986.

[26] Crepin J, Johnson RL. Soil sampling for environmental assessment. (Chapter 2). In: Carter MR, ed. Soil Sampling and Methods of Analysis. Ann Arbor, MI, USA: Canadian Society of Soil Science, Lewis Publishers, 1993.

[27] Warrick AW, Myers DE, Nielsen DR. Geostatistical methods applied to soil science. In: Klute A, ed. Methods of Soil Analysis. Part 1. Physical and Mineralogical Methods. 2nd ed. Madison, WI, USA: Soil Science society of America, Inc, 1986.

[28] Hesse PR. A Textbook of Soil Chemical Analysis. New York, NY, USA: Chemical Publishing Co, Inc, 1971.

[29] Bates TE. Soil handling and preparation. In: M.R. Carter, ed. Soil Sampling and Methods of Analysis. Boca Raton, FL, USA: (Canadian Society of Soil Science) Lewis Publishers, Inc, 1993: 19-24.

[30] Jackson ML. Soil Chemical Analysis. Englewood Cliffs, NJ, USA: Prentice-Hall, Inc, 1958.

[31] Soil Survey Staff. Keys to Soil Taxonomy. 12th ed. US Department of Agriculture, Natural Resources Conservation Service, 2014. (Accessed July 31, 2014, at http://www.nrcs.usda.gov/wps/portal/nrcs/detail/soils/survey/class/?cid= nrcs142p2_053580)

[32] Freese F. Elementary Forest Sampling. Agriculture Handbook No. 232. U.S. Department of Agriculture. Corvallis, OR, USA: Oregon State University (Reprinted 1981), 1962.

[33] Wilde SA, Corey RB, Iyer JG, Voigt GK. Soil and Plant Analysis for Tree Culture. New Delhi, India: Oxford & IBH Publishing Co, 1979.

[34] Tucker TC. Diagnosis of nitrogen deficiency in plants. In: Hauck RD, ed. Nitrogen in Crop Production. Madison, WI, USA: American Society of Agronomy, 1984: 249-262.

[35] Cook RL, Ellis BG. Soil Management—A World View of Conservation and Production. New York, NY, USA: John Wiley & Sons, 1987.

[36] Leaf AL. Plant analysis as an aid in fertilizing forests. In: Walsh LM, Beaton JD, eds. Soil Testing and Plant Analysis. Madison, WI, USA: Soil Science Society of America, 1973: 427-454.

[37] Leaf AL, Bickelhaupt DH. Possible mutual prediction between black cherry and sugar maple foliar analysis data. Soil Sci Soc Am Proc 1975; 39: 983-985.

[38] Munson RD, Nelson WL. Principles and practices in plant analysis. In: Walsh LM, Beaton JD, eds. Soil Testing and Plant Analysis. Madison, WI, USA: Soil Science Society of America, 1973: 223-248.

[39] Strickland TC, Sollins P. Improved method for separating light- and heavy-fraction organic material from soil. Soil Sci Soc Am J 1987; 51: 1390-1393.

[40] Ulrich A, Ririe D, Hills FJ, George AG, Moore MD, Johnson CM. Plant Analysis—A guide for Sugar Beet Fertilization. Berkeley, CA, USA: Calif Agr Exp Sta Bul, 1959: 766.

[41] Leggett GE, Westerman DT. Determination of mineral elements in plant tissue using trichloroacetic acid extraction. J Agric Food Chem 1973; 21: 65-69.

[42] Gardner WH. Water content. In: Klute A, ed. Methods of Soil analysis. Part 1. Physical and Mineralogical Methods. 2nd Ed. Madison, WI, USA: American Society of Agronomy, Soil Science Society of America, 1986: 493-544.

[43] Hook WR, Livingston NJ, Sun ZJ, Hook PB. Remote diode shorting improves measurement of soil water by time domain reflectometry. Soil Sci Soc Am J 1992; 56: 1384-1391.

[44] Federer CA. Nitrogen mineralization and nitrification: Depth variation in four New England forest soils. Soil Sci Soc Am J 1983; 47: 1008-1014.

[45] Gee GW, Bauder JW. Particle-size analysis. In: Klute A, ed. Methods of Soil Analysis, Part 1. Physical and Mineralogical Methods. 2nd ed. Madison, WI, USA: American Society of Agronomy, Soil Science Society of America, 1986: 383-411.

[46] USDA Soil Survey Field and Laboratory Methods Manual. Soil Survey Investigations Report No. 51 Version 2. 2014. (Accessed July 27, 2014, at http://www.nrcs.usda.gov/wps/portal/nrcs/detail/soils/survey/publication/?cid=stelprdb 1247805).

[47] Schoeneberger PJ, Wysocki DA, Benham EC, Broderson WD. Field book for describing and sampling soils. National Soil Survey Center, Lincoln, NE, USA: Natural Resources Conservation Service, USDA, 1998.

[48] Baver LD, Gardner WH, Gardner WR. Soil Physics. 4th ed. New York, NY, USA: John Wiley & Sons, Inc, 1972: 284-286.

[49] Sheldrick BH, Wang C. Particle size distribution. In: Carter MR, ed. Soil Sampling and Methods of Analysis. Ann Arbor, MI, USA: (Canadian Society of Soil Science) Lewis Publishers, 1993: 499-511.

[50] Food and Agriculture Organization (FAO) of the United Nations. Soil Texture. (Accessed July 27, 2014, at ftp://ftp.fao.org/fi/cdrom/fao_training/FAO_ Training/General/x6706e/x6706e06.htm)

[51] Day PR. Particle fractionation and particle-size analysis. In: Black CA, eds. Methods of Soil Analysis, Part I. Madison, WI, USA: American Society of Agronomy, 1965: 545-567.

[52] Kunze GW. Pretreatment for mineralogical analysis. In: Black CA, ed. Methods of Soil Analysis, Part 1. Madison, WI, USA: American Society of Agronomy, 1965: 568-577.

[53] Baver LD, Gardner WH, Gardner WR. Soil Physics. 4th ed. New York, NY, USA: John Wiley & Sons, Inc, 1972.

[54] Kozlowski TT. Water Deficits and Plant Growth. Vol. I. New York, NY, USA: Academic Press, 1968.

[55] Jones C, Jacobsen J. Plant Nutrition and Soil Fertility. Nutrient Management Module No. 2. Bozeman, MT, USA: Montana State University Extension Service Continuing Education Series, Report No. 4449-2, June 2005.

[56] Jacobsen J, Lorbeer S. Soil, Plant and Water Analytical Laboratories for Montana Agriculture. Bozeman, MT, USA: Montana State University Extension, 1998: EB 150. (Accessed July 28, 2013, at http://animalrangeextension. montana.edu/attachments/dennis/EB0150.pdf)

[57] Lawrence G, Vogt KA, David M, et al. Atmospheric deposition effects on surface waters, soils, and forest productivity in northeastern United States—Advances since NAPAP. In: Mickler R, Birdsey R, Hom J, eds. Responses of Northern Forests to Environmental Change. New York, NY, USA: Springer-Verlag, 1999.

[58] Wargo PM, Vogt KA, Vogt DJ, et al. Vitality and chemistry of forest floor roots in red spruce, *Picea rubens* Sarg. Dominated stands characterized by a gradient of soil Al/Ca ratios in the northeastern United States. Can J For Res, 2003; 33: 635-652.

[59] Thomas GW. Problems encountered in soil testing methods. In: Soil Testing and Plant Analysis, Part 1. Madison, WI, USA: Soil Science Society of America, 1967: 37-54.

[60] McLean EO. Soil pH and lime requirement. In: Page AL, ed. Methods of Soil Analysis, Part 2: Chemical and Microbiological Properties. 2nd ed. Madison,

WI, USA: American Society of Agronomy, Soil Science Society of America, 1982: 199-224.

[61] Coleman NT, Thomas GW. The basic chemistry of soil acidity. In: Pearson RW, Adams F, eds. Soil Acidity and Liming. Madison, WI, USA: American Society of Agronomy, Soil Science Society of America, 1967; 12: 1-41.

[62] Westcott CC. pH Measurements. San Diego, CA, USA: Academic Press, Inc, 1978.

[63] Willard HH, Merritt LL, Dean JA, Settle FA. Instrumental Methods of Analysis. 7th ed. Belmont, CA, USA: Wadsworth Publishing Co, 1988.

[64] pH glass electrode. Wikipedia—the Free Encyclopedia. (Accessed July 28, 2013, at http://en.wikipedia.org/wiki/Glass_electrode)

[65] Soil Survey Division Staff, eds. Soil Survey Manual. Ch 3. Examination and Description of Soils. United States Department of Agriculture Handbook No. 18. Washington, DC, USA: US Dept of Agric, 1993. (Accessed January 22, 2014, at http://www.nrcs.usda.gov/wps/portal/nrcs/detail/soils/survey/?cid=nrcs142p2 _054262)

[66] The salt of the earth. Bible (King James Version), Matthew 5: 13.

[67] Rhoades JD. Soluble Salts. In: Page AL, ed. Methods of Soil Analysis, Part 2: Chemical and Microbiological Properties. 2nd ed. Madison, WI, USA: American Society of Agronomy, Soil Science Society of America, 1982: 167-179.

[68] Geraldson CM. Nutrient intensity and balance. In: Stelly M, ed-in-chief. Soil Testing: Correlating and Interpreting the Analytical Results. Madison, WI, USA: American Society of Agronomy, Crop Science Society of America, and Soil Science Society of America, 1977: 75-84.

[69] Richards LA. Diagnosis and improvement of saline and alkali soils. In: US Salinity Laboratory Staff, ed. US Dept of Agriculture Handbook No. 60, 1954. (Accessed July 29, 2013, at http://www.ars.usda.gov/Services/docs.htm?docid =10158).

[70] Geraldson CM. Soil solution soluble salts as an indicator of fertility level and nutrient balance. Soil Sci Soc Florida Proc 1955; 15: 22-30.

[71] Geraldson CM. Soil soluble salts-determination of and association with plant growth. Fla State Hortic Soc 1957; 70: 121-126.

[72] Kelley WP. Cation Exchange in Soils. New York, NY, USA: Reinhold Publishing Corp, 1948.

[73] Sposito G. The Chemistry of Soils. New York, NY, USA: Oxford University Press, 1989.

[74] Gillman GP, Sumpter EA. Modification to the compulsive exchange method for measuring exchange characteristics of soils. Aust J Soil Res 1986; 24: 61-66.

[75] Polemio M, Rhoades JD. Determining cation exchange capacity: A new procedure for calcareous and gypsiferous soils. Soil Sci Soc Am J 1977; 41: 524-528.

[76] Gillman GP. A proposed method for the measurement of exchange properties of highly weathered soils. Aust J Soil Res 1979; 17: 129-139.

[77] Sumner ME, Miller WP. Cation exchange capacity and exchange coefficients. In: Sparks DL, ed. Methods of Soil Analysis. Part 3. Chemical Method. Madison, WI, USA, Soil Science Society of America, American Society of Agronomy, 1996; 5: 1201-1229.

[78] Chapman HD. Cation exchange capacity. In: Black CA, ed. Methods of Soil Analysis. Madison, WI, USA: American Society of Agronomy, 1965; 9: 891-901.

[79] Rhoades JD, Krueger DB. Extraction of cations from silicate minerals during the determination of exchangeable cations in soils. Soil Sci Soc Am Proc, 1968; 32: 488-492.

[80] Bower C. Fixation of ammonium in difficultly exchangeable forms under moist conditions by some soils of semi-arid regions. Soil Sci 1950; 70: 375-383.

[81] Bohn HL, McNeal BL, O'Connor GA. Soil Chemistry. New York, NY, USA: John Wiley & Sons, 1979.

[82] Suarez DL. Beryllium, magnesium, calcium, strontium, and barium. In: Sparks DL, ed. Methods of Soil Analysis. Part 3. Chemical Method. Madison, WI, USA: Soil Science Society of America, American Society of Agronomy, 1996; 5: 575-601.

[83] Thomas GW. Exchangeable cations. In: Page AL, ed. Methods of soil analysis, Part 2. Chemical and Microbiological Properties, 2nd ed. Madison, WI, USA: Soil Science Society of America, 1982: 159-165.

[84] Bertsch PM, Bloom PR. Aluminum. In: Sparks DL, ed. Methods of Soil Analysis. Part 3. Chemical Methods. Madison, WI, USA: Soil Science Society of America, American Society of Agronomy, 1996: 517-550.

[85] Mader D. A laboratory manual for forest soils. Amherst, MA, USA: University of Massachusetts at Amherst, 1975.

[86] Keeney DR, Nelson DW. Nitrogen in organic forms. In: Page AL, ed. Methods of Soil Analysis, Part 2. Chemical and Microbiological Properties. 2nd ed. Madison, WI, USA: American Society of Agronomy, Soil Science Society of America, 1982: 643-698.

[87] Young JL, Aldag RW. Inorganic forms of nitrogen in soil. In: Stevenson FJ, ed. Nitrogen in Agricultural Soils. Madison, WI, USA: American Society of Agronomy, Soil Science Society of America, 1982: 43-66.

[88] Maynard DG, Kalra YP. Nitrate and Exchangeable Ammonium Nitrogen. In: Carter MR, ed. Soil Sampling and Methods of Analysis. Ann Arbor, MI, USA: (Canadian Society of Soil Science) Lewis Publishers, 1993: 25-38.

[89] Haynes RJ, Swift RS. Effect of rewetting air-dried soils on pH and accumulation of mineral nitrogen. J Soil Sci 1989; 40: 340-347.

[90] Olsen SR, Sommers LE. Phosphorus. In: Page AL, ed. Methods of Soil Analysis, Part 2. Chemical and Microbiological Properties. 2nd ed. Madison, WI, USA: American Society of Agronomy, Soil Science Society of America, 1982: 403-430.

[91] Cathcart JB. World phosphate reserves and resources. In: Khasawneh FE, Sample EC, Kamprath EJ, eds. The Role of Phosphorus in Agriculture. Madison, WI, USA: Am Soc of Agron, Crop Sci Soc of Am, Soil Sci Soc of Am, 1980; 1-18.

[92] McClellan GH, Gremillion LR. Evaluation of phosphatic raw materials. In: Khasawneh FE, Sample EC, Kamprath EJ, eds. The Role of Phosphorus in Agriculture. Madison, WI, USA: Am Soc of Agron, Crop Sci Soc of Am, Soil Sci Soc of Am, 1980: 43-80.

[93] Kuo S. Phosphorus. In: Sparks DL, ed. Methods of Soil Analysis. Part 3. Chemical Methods. Madison, WI, USA: Soil Science Society of America, American Society of Agronomy, 1996: 869-919.

[94] Olsen SR, Cole CV, Watanabe FS, Dean LA. Estimation of available phosphorus in soils by extraction with sodium bicarbonate. U.S. Department of Agriculture Circular 939, 1954.

[95] Bray RH, Kurtz LT. Determination of total, organic and available forms of phosphorus in soils. Soil Sci 1945; 59: 39-45.

[96] Egnér H, Riehm H, Domingo WR. Untersuchungen über die chemische Bodenanalyse als Grundlage für die Beurteilung des Nährstoffzustandes der Böden. 2. Chemische Extraktionsmethoden zur Phosphor- und Kaliumbestimmung. Uppsala, Sweden: Kungl Landbrukshögskolans Annaler 1960; 26: 199-215.

[97] Banderis AS, Barter DH, Henderson K. The use of polyacrylamide to replace carbon in the determination of Olsen's extractable phosphate in soil. J Soil Sci 1976; 27: 71-74.

[98] Nelson, WL, Mehlich A, Winters E. The development, evaluation, and use of soil tests for phosphorus availability. Agronomy 1953; 4: 153-188.

[99] Dalal RC, Hallsworth EG. Evaluation of the parameters of soil phosphorus availability factors in predicting yield response and phosphorus uptake. Soil Sci Soc Am J 1976; 40: 541-546.

[100] Schoenau JJ, Huang WZ. Anion-exchange membrane, water, and sodium bicarbonate extractions as soil tests for phosphorus. Commun Soil Sci Plant Anal 1991; 22: 465-492.

[101] Saggar, S, MJ Hedley and RE White. A simplified resin membrane technique for extracting phosphorus from soils. Fertil Res 1990; 24: 173-180.

[102] Bowman RA, Olsen SR. A reevaluation of phosphorus-32 and resin methods in a calcareous soil. Soil Sci Soc Am J 1979; 43: 121-124.

[103] Gunary D, Sutton CD. Soil factors affecting plant uptake of phosphate. J Soil Sci 1967; 18: 167-173.

[104] Olsen SR, Watanabe FS. Diffusive supply of phosphorus in relation to soil textural variations. Soil Sci 1970; 110: 318-327.

[105] Amer F, Bouldin DR, Black CA, Duke FR. Characterisation of soil phosphorus by anion exchange resin adsorption and ^{32}P equilibration. Plant Soil 1955; 6: 391-408.

[106] Talibudeen O. Isotopically exchangeable phosphorus in soils, II. J Soil Sci 1957; 8: 86-96.

[107] Russell, RS, EW Russell and PG Marais. Factors affecting the ability of plants to absorb phosphate from soils. I. J. Soil Sci 1957; 8: 248-267.

[108] SSL Method 6A4. NRCS. Soil Survey Laboratory Methods Manual. Soil Survey Investigations Report No. 42, Version 3.0. Washington, DC, USA: US Department of Agriculture, Natural Resources Conservation Service, National Soil Survey Center, 1996.

[109] Schlesinger WH. Biogeochemistry: An Analysis of Global Change, 2nd ed. San Diego, CA, USA: Academic Press, 1991.

[110] Brown S, Lugo AE, Iverson LR. Processes and lands for sequestering carbon in the tropical forest landscape. Water Air Soil Poll 1992; 64: 139-155.

[111] Huntington TG. Carbon sequestration in an aggrading forest ecosystem in southeastern USA. Soil Sci Soc of Am J 1995; 59: 1459-1467.

[112] Nelson DW, Sommers LE. Total carbon, organic carbon, and organic matter. In: Sparks DL, ed. Methods of Soil Analysis. Part 3. Chemical Methods. Madison, WI, USA: Soil Science Society of America, American Society of Agronomy, 1996: 961-1010.

[113] Walkley A, Black IA. An examination of the Degtjareff method for determining soil organic matter and a proposed modification of the chromic acid titration method. Soil Sci 1934; 37: 29-38.

[114] Manahan SE. Environmental Chemistry. Boca Raton, FL, USA: Lewis Publishers, 1999.

[115] Johnson DW, Todd DE. Relationships among iron, aluminum, carbon, and sulfate in a variety of forest soils. Soil Science Society of America Journal, 1983; 47: 792-426.

[116] Ugolini FC, Minden R, Dawson H, Zachara J. An example of soil processes in the Abies amabilis zone of central Cascades, Washington. Soil Science 1977; 124: 291-302.

[117] Johnson DW, Cole DW. Anion mobility in soils: Relevance to nutrient transport from terrestrial ecosystems. Environ Int 1980; 3: 79-90.

[118] Van Cleve K, Chapin FS III, Dyrness CT, Viereck LA. Element cycling in taiga forests: State-factor control. BioScience 1991; 41: 78-88.

[119] Kulmatiski A, Vogt KA, Vogt DJ, et al. Northeastern US forest response to cation remediation. Canadian Journal Forest Research 2007; 37: 1574-1585.

[120] Bormann FH, Likens GE. Pattern and Process in a Forested Ecosystem— Disturbance, Development and the Steady State Based on the Hubbard Brook Ecosystem Study. New York, NY, USA: Springer-Verlag, 1979.

[121] Shortle WC, Smith KT. Aluminum-induced calcium deficiency syndrome in declining red spruce. Science 1988; 240: 1017-1018.

[122] Lindsay WL, Moreno EC. Phosphate phase equilibria in soils. Madison, WI, USA: Soil Science Society of America Proceedings 1960; 24: 177-182.

[123] Ross S. Soil Process—A Systematic Approach. New York, NY, USA: Routledge, 1989.

[124] Tilley J. Research in the Greeley Analytical Laboratory. Unpublished data. Yale School of Forestry and Environmental Studies, Greeley Analytical Lab, 1990-2000.

[125] Kjeldahl J. Neue Methode zur Bestimmung des Stickstoffs in organischen Körpern. Z Anal Chem 1883; 22: 366-382.

[126] Bremner JM. Nitrogen—total. In: Sparks DL, ed. Methods of Soil Analysis, Part 3: Chemical Methods. Madison, WI, USA: Soil Science Society of America, American Society of Agronomy, 1996: 1085-1121.

[127] Parkinson JA, Allen SE. A wet oxidation procedure suitable for the determination of nitrogen and mineral nutrients in biological material. Communications in Soil Science and Plant Analysis 1975; 6(1): 1-11.

[128] Dumas JBA. Procédés de l'analyse organique. Ann Chim Phys 1831; 247: 198-213.

[129] McGill WB, Figueiredo CT. Total Nitrogen. In: Carter MR, ed. Soil Sampling and Methods of Analysis. Ann Arbor, MI, USA: (Canadian Society of Soil Science) Lewis Publishers, 1993: 201-211.

[130] Bellomonte G, Costantini A, Giammarioli S. Comparison of modified automatic Dumas method and the traditional Kjeldahl method for nitrogen determination in infant food. J Assoc Offic Anal Chem 1987; 70: 227-229.

[131] Siccama TG, Johnson AH. Personal communication. Yale University and University of Pennsylvania, 1990.

[132] APHA (American Public Health Association). Standard Methods for the Examination of Water and Wastewater. 14th ed. Washington, DC, USA: American Public Health Association, 1976.

[133] D'Elia CF, Steudler PA, Corwin N. Determination of total nitrogen in aqueous samples using persulfate digestion. Limnol Oceanogr 1977; 22: 760-764.

[134] Weaver RW, Angle S, Bottomley P, eds. Methods of Soil Analysis. Part 2. Microbiological and Biochemical Properties. Madison, WI, USA: Soil Science Society of America, 1994.

[135] Robertson GP, Coleman DC, Bledsoe CS, Sollins P, eds. Standard Soil Methods for Long-Term Ecological Research. Long-Term Ecological Research Network Series. New York, NY, USA: Oxford University Press, 1999.

[136] Kirk JL Beaudette LA, Hart M, et al. Methods of studying soil microbial diversity. J of Microbiological Methods 2004; 58: 169-188.

[137] Schinner FR, Öhlinger R, Kandeler E, Margesin R, eds. Methods in Soil Biology. New York, NY, USA: Springer, 2013.

[138] Bergey's Manual of Systematic Bacteriology, 1st ed.at Earthlife.net., 2013. (Accessed August 2, 2013, at http://www.earthlife.net/prokaryotes/phyla.html)

[139] Agrios GN. Plant Pathology. 5th ed. New York, NY, USA; Elsevier Academic Press, 2005.

[140] Paul EA, Harris D, Klug MJ, Ruess RW. The determination of microbial biomass. In: Robertson GP, Coleman DC, Bledsoe CS, Sollins P, eds. Standard Soil Methods for Long-Term Ecological Research. Long-Term Ecological Research Network Series. New York, NY, USA: Oxford University Press, 1999: 291-317.

[141] Brundrett M. Mycorrhizal associations: The web resource. 2008. (Accessed June 21, 2013, at http://mycorrhizas.info/)

[142] Smith SE, Read DJ. Mycorrhizal Symbiosis, 3rd ed. New York, NY, USA: Academic Press, 2008.

[143] Johnson NC, O'Dell TE, Bledsoe CS. Methods for ecological studies of mycorrhizae. In: Robertson GP, Coleman DC, Bledsoe CS, Sollins P, eds. Standard Soil Methods for Long-Term Ecological Research. Long-Term Ecological Research Network Series. New York, NY, USA: Oxford University Press, 1999: 378-412.

[144] Schenck NC, ed. Methods and Principles of Mycorrhizal Research. St. Paul, MN, USA: American Phytopathological Society, 1982.

[145] Norris JR, Read DJ, Varma AK, eds. Methods in Microbiology. Volume 23, Techniques for the Study of Mycorrhiza. London, UK: Academic Press, 1991.

[146] Norris JR, Read DJ, Varma AK, eds. Methods in Microbiology. Volume 24, Techniques for the Study of Mycorrhiza. London, UK: Academic Press, 1992.

[147] Brundrett M, Melville L, Peterson L, eds. Practical Methods in Mycorrhiza Research. Sidney, BC, Canada: Mycologue Publications, 1994.

[148] Brundrett MC, Bougher N, Dell B, Grove T, Malajczuk N. Working with Mycorrhizal Fungi in Forestry and Agriculture. Canberra, Australia: Australian Center for International Agricultural Research, 1996.

[149] Brundrett M. Mycorrhizal associations: The web resource section 10, methods for identifying mycorrhizas. 2008. (Accessed June 21, 2013, at http://mycorrhizas.info/method.html)

[150] Goodman DM, Durall DM, Trofymow JA, Berch SM. A Manual of Concise Descriptions of North American Ectomycorrhizae. Sidney, BC, Canada: Mycologue Publications, 1996.

[151] INVAM. The international culture collection of (vesicular) arbuscular mycorrhizal fungi (INVAM), 2013. (Accessed June 22, 2012, at http://invam.caf.wvu.edu/)

[152] Beg. The International Bank for the Glomeromycota (beg), 2013. (Accessed June 20, 2013, at http://www.i-beg.eu/)

[153] Kormanik PP, McGraw AC. Quantification of vesicular-arbuscular mycorrhizae in plant roots. In: Schenck NC, ed. Methods and Principles of Mycorrhizal Research. St. Paul, Minnesota, USA: American Phytopathological Society, 1982: 37-45.

[154] Allen MF, Moore TS Jr, Christensen M, Stanton N. Growth of vesicular arbuscular mycorrhizal and non-mycorrhizal Bouteloua gracilis in a defined medium. Mycologia 1979; 71: 666-669.

[155] O'Dell TE, Luoma DL, Molina RJ. Ectomycorrhizal fungal communities in young, managed and old growth Douglas-fir stands. Northwest Environmental Journal 1992; 8: 166-168.

[156] Agerer R. Colour Atlas of Ectomycorrhizae. 1st-5th delivery. Schwabisch-Gmund, Germany: Einhorn-Verlag, 1987-1991.

[157] Agerer R. Characterization of ectomycorrhiza. In: Norris JR, Read DJ, Varma AK, eds. Methods in Microbiology. Volume 23. Techniques for the Study of Mycorrhiza. London, UK: Academic Press, 1991: 25-27.

[158] Franson RL, Bethlenfalvay GJ. Infection unit method of vesicular-arbuscular mycorrhizal propagule determination. Soil Science Society of America Journal 1989; 53: 754-756.

[159] Jasper DA, Abbott LK, Robson AD. Soil disturbance reduces the infectivity of external hyphae of vesicular-arbuscular mycorrhizal fungi. New Phytologist 1989; 112: 93-99.

[160] Perry DA, Amaranthus MP, Borchers JG, Bouchers SL, Brainerd RE. Bootstrapping in ecosystems. BioScience 1989; 39: 230-237.

[161] Brundrett MC, Abbott LK. Mycorrhizal fungus propagules in the Jarrah Forest. 1. Seasonal study of inoculum levels. New Phytologist 1994; 127: 539-546.

[162] Morton JB, Bentivenga SP, Wheeler WW. Germ plasm in the international collection of arbuscular and vesicular-arbuscular mycorrhizal fungi (INVAM) and procedures for culture development, documentation and storage. Mycotaxon 1993; 48: 491-528.

[163] Morton JB, Bentivenga SP, Bever JD. Discovery, measurement, and Interpretation of diversity in arbuscular endomycorrhizal fungi (Glomales, Zygomycetes). Canadian Journal of Botany 1995; 73: S25-S32.

[164] Cline ET Ammirati JF, Edmonds RL. Does proximity to mature trees influence ectomycorrhizal fungus communities of Douglas-fir seedlings. New Phytologist 2005; 166: 993-1009.

[165] Gardes M, Bruns TD. ITS primers with enhanced specificity for basidiomycetes—application to the identification of mycorrhizae and rusts. Molecular Ecology 1993; 2: 113-118.

[166] White TJ, Bruns TD, Lee SB, Taylor JW. Amplification and direct sequencing of fungal ribosomal RNA genes for phyolgenetics. In: Innis MA, Gelfand DH, Sninsky JJ, and White TJ, eds. PCR Protocols: A guide to Methods and Applications. New York, NY, USA: Academic Press, 1990: 315-322.

[167] Moncalvo J-M, Lutzoni FM, Rehner SA, Johnson J, Vilgalys R. Phylogenetic relationships of agaric fungi based on nuclear large subunit ribosomal DNA sequences. Systematic Biology 2000; 49: 278-305.

[168] National Center for Biotechnology Information web-based BLAST search engine, 2013. (Accessed August 3, 2013, at http://blast.ncbi.nlm.nih.gov/Blast.cgi)

[169] Vogt KA, Grier CC, Meier CE, Edmonds RL. Mycorrhizal role in net primary production and nutrient cycling in Abies amabilis ecosystems in western Washington. Ecology 1982; 63: 370-380.

[170] Gardner JH Malajczuk N. Recolonisation of rehabilitated bauxite mine sites in Western Australia by mycorrhizal fungi. Forest Ecol Management 1988; 24: 27-42.

[171] Holland EA, Robertson GP, Greenberg J, Groffman PM, Boone RD, Gosz JR. Soil CO_2, N_2, and CH_4 exchange. In: Robertson GP, Coleman DC, Bledsoe CS, and Sollins P, eds. Standard Soil Methods for Long-Term Ecological Research. Long-Term Ecological Research Network Series. New York, NY, USA: Oxford University Press, 1999: 185-201.

[172] IRGA. LI-COR, Inc. Environmental and Biotechnology Research Systems. Lincoln, NE, USA, 2013. (Accessed August 12, 2013, at http://www.licor.com)

[173] Grogan P. CO_2 flux measurement using soda lime: correction for water formed during CO_2 adsorption. Ecology 1998; 79(4): 1467-1468.

[174] Edwards NT. The use of soda lime for measuring respiration rates in terrestrial systems. Pedobiologia 1982; 23: 321-330.

[175] Harmon ME, Nadelhoffer KJ, Blair JM. Measuring decomposition, nutrient turnover, and stores in plant litter. In: Robertson GP, Coleman DC, Bledsoe CS, Sollins P, eds. Standard Soil Methods for Long-Term Ecological Research. Long-Term Ecological Research Network Series. New York, NY, USA: Oxford University Press, 1999: 202-240.

[176] Olson JS. Energy stores and the balance of producers and decomposers in ecological systems. Ecology 1963; 44: 322-331.

[177] Hurlbert SH. Pseudoreplication and the design of ecological field experiments. Ecological Monographs 1984; 54(2): 187-211.

[178] Harmon ME, Sexton J. Guidelines for Measurements of Woody Detritus in Forest Ecosystems. US LTER Network Office Publication, No. 20. Seattle, WA, USA: University of Washington, 1996.

[179] Harmon ME, Franklin JF, Swanson F, et al. Ecology of coarse woody debris in temperate ecosystem. Advances in Ecological Research 1986; 15: 133-302.

[180] Harmon ME, Cromack K Jr, Smith BG. Coarse woody debris in mixed conifer forests, Sequoia National Park, California. Canadian Journal of Forest Research 1987; 17: 1265-1272.

[181] Harmon ME. Long-Term Experiments on Log Decomposition at the H. J. Andrews Experimental Forest. General Technical Report PNW-280. Portland, OR, USA: USDA Forest Service, 1992.

[182] Currie WS, Harmon ME, Burke, IC, Hart SC, Parton, WJ, Silver W. Cross-biome transplants of plant litter show decomposition models extend to a broader climatic scale but lose predictability at the decadal time scale. Global Change Biology 2010; 16: 1744-1761.

[183] Smith, AC, Bhatti JS, Chen H, Harmon ME, Arp PA. Modelling above- and below-ground mass loss and N dynamics in wooden dowels (LIDET) placed across North and Central America biomes at the decadal time scale. Ecological Modelling 2011; 222: 2276-2290.

[184] Wieder RR, GE Lang. A critique of the analytical methods used in examining decomposition data obtained from litter bags. Ecology 1982; 63: 1636-1642.

[185] Edmonds RL. Long-term decomposition and nutrient dynamics in Pacific silver needles in western Washington, USA. Canadian Journal of Forest Research 1984; 14: 395-400.

[186] Sinsabaugh RL, Klug MJ, Collins HP, Yeager PE, Peterson SO. Characterizing soil microbial communities. In: Robertson GP, Coleman DC, Bledsoe CS, Sollins P, eds. Standard Soil Methods for Long-Term Ecological Research. Long-Term Ecological Research Network Series. New York, NY, USA: Oxford University Press 1999; 318-348.

[187] Dick RP. Methods of Soil Enzymology. SSSA Book Series No. 9, Madison, WI, USA: Soil Science Society of America, 2011.

[188] Burns RG. Soil Enzymes. Academic Press, New York, NY, USA, 1978.

[189] Ho I. Acid phosphatase activity in forest soil. Forest Sci 1979; 25: 567-568.

[190] Saiya-Cork KR, Sinsabaugh RL, Zak DR. Effects of long term nitrogen deposition on extracellular enzyme activity in an Acer saccharum forest soil. Soil Biology and Biochemistry 2002; 34: 1309-1315.

[191] Biolog, Inc., 2013. (Accessed August 2, 2013, at http://www.biolog.com)

[192] Yeates C, Gillins MR, Davison AD, Altavilla N, Veal DA. Methods for microbial DNA extraction from soil for PCR amplification. Biol Proced Online, May 1998-April 1999; 1: 40-47. (Accessed June 22, 2013, at http://www.ncbi.nlm.nih.gov/pmc/articles/PMC140122/)

[193] SoilMaster DNA Extraction Kit, 2013. (Accessed August 2, 2013, at http://www.epibio.com/ products/ nucleic-acid -purification -extraction-kits/dna-extraction/soilmaster-dna-extraction-kit).

[194] SurePrep™ Soil DNA Isolation Kit, 2013. (Accessed August 2, 2013, at http://fscimage.fishersci.com/cmsassets/downloads/segment/Scientific/pdf/Chemicals/SurePrep_SoilDNA_techmanual.pdf)

[195] PowerSoil, PowerMax and UltraClean soil DNA isolation kits, 2013. (Accessed August 2, 2013, at http://www.mobio.com/soil-dna-isolation/)

[196] NucleoSpin Soil DNA Isolation Kit, 2013. (Accessed August 2, 2013, at http://www.clontech.com/US/Support/Applications/Nucleic_Acid_Purification/DNA_From_Soil?gclid=CPXt27D367UCFadFMgodyBgA0w)

[197] FastDNA Spin Kit, 2013. (Accessed August 2, 2013, at http://images.www.mpbio.com/docs/fastprep/FastDNASpinKitforsoil.pdf)

[198] Sanger F. Sanger method of DNA sequencing, 2013. (Accessed August 3, 2013, at http://en.wikipedia.org/wiki/Sanger_sequencing)

[199] Roesch LFW, Fulthorpe RR, Riva A, et al. Pyrosequencing enumerates and contrasts soil microbial diversity. The International Soc for Microbial Ecology (ISME) Journal 2007; 1: 283-290.

[200] Urich T, Lanzén A, Qi J, Huson DH, Schleper C, Schuster SC. Simultaneous assessment of soil microbial community structure and function through analysis of the meta-transcriptome. PLoS ONE, 2008;3(6):e2527.doi:10.1371/hiyrbak, oibe,0002527. (Accessed June 22, 2013, at http://www.plosone.org/article/fetch Object?uri=info%3Adoi%2F10.1371%2Fjournal.pone.0002527&representation=PDF)

[201] Coleman DC, Blair JM, Elliott ET, Wall DH. Soil invertebrates. In: Robertson GP, Coleman DC, Bledsoe CS, Sollins P, eds. Standard Soil Methods for Long-Term Ecological Research. Long-Term Ecological Research Network Series. New York, NY, USA: Oxford University Press, 1999; 349-377.

[202] Brian MV. Social Insect Populations. London, UK: Academic Press, 1965.

[203] Lee KE, Wood TG. Termites and Soils. London, UK: Academic Press, 1971.

[204] Dindal DL. Soil Biology Guide. New York, NY, USA: Wiley, 1990.

[205] Hendrix PF. Earthworm Ecology and Biogeography in North America. Boca Raton, FL, USA: Lewis Publishers, 1995.

[206] Hendrix PF, Mueller BR, Bruce RR, Langdale GR, Parmelee RW. Abundance and distribution of earthworms in relation to landscape factors on the Georgia Piedmont, U.S.A. Soil Biology and Biochemistry 1992; 24: 1357-1361.

[207] Geurs M, Bongers J, Brussard L. Improvements of the heptanes flotation method for collecting microarthropods from silt loam soil. Agriculture, Ecosystems and Environment 1991; 34: 213-221.

[208] Moldenke AR. Arthropods. In: Weaver RW, Angle S, Bottomley P, eds. Methods of Soil Analysis. Part 2, Microbiological and Biochemical Properties. Madison, WI, USA: Soil Science Society of America, 1994: 517-542.

[209] Tullgren funnel. A schematic of the Tullgren funnel to extract soil invertebrates from the soil. Wikipedia Commons, 2013. (Accessed August 12, 2013, at http://en.wikipedia.org/wiki/Tullgren_funnel)

[210] Robertson, GP, Wedin, D, Groffman, PM, Blair, JM, Holland, EA, Nadelhoffer, KJ, Harris, D. Soil carbon and nitrogen availability: Nitrogen mineralization, nitrification, and soil respiration potentials. In: Robertson GP, Coleman DC, Bledsoe CS, Sollins P, eds. Standard Soil Methods for Long-Term Ecological Research. Long-Term Ecological Research Network Series. New York, NY, USA: Oxford University Press, 1999; 258-271.

[211] Norton JM, Stark, JM. Regulation and measurement of nitrification in terrestrial systems. In Klotz, MG (Ed.): Methods in Enzymology. Research on Nitrification and Related Processes. Burlington, MA, Academic Press, 2011; 343-368.

[212] Vitousek,PM, Menge, DNL, Reed SC, Cleveland, CC. Biological nitrogen fixation: Rates, patterns and ecological controls in terrestrial systems. Philosophical Transactions off the Royal Society B 2013, 368: 20130119.

[213] Myrold, DD, Ruess, RW, Klug, MJ. Dinitrogen fixation. In: Robertson GP, Coleman DC, Bledsoe CS, Sollins P, eds. Standard Soil Methods for Long-Term Ecological Research. Long-Term Ecological Research Network Series. New York, NY, USA: Oxford University Press, 1999; 241-257.

[214] Groffman, PM, Holland, EA, Myrold, DD, Robertson, GP, Zou, X. Denitrification. In: Robertson GP, Coleman DC, Bledsoe CS, Sollins P, eds. Standard Soil Methods for Long-Term Ecological Research. Long-Term Ecological Research Network Series. New York, NY, USA: Oxford University Press, 1999; 272-288.

[215] Groffman, PM, Altabet, MA,. Böhlke, JK, Butterbach-Bahl,K, David, MB, Firestone, MK, Giblin, AE, Kana, TM, Nielsen, LP, Voytek, MA. Methods for measuring denitrification: Diverse approaches to a difficult problem. Ecological Applications 2006, 16: 2091-2122.

[216] Klesta EJ Jr, Bartz JK. Quality assurance and quality control. In: Sparks DL, ed. Methods of soil analysis. Part 3. SSSA Book Series No. 5, Madison, WI, USA: Soil Science Society of America, 1996.

[217] Quality Assurance of Chemical Measurements. Boca Raton, FL, USA. Lewis Publishers, 1987.

Subject Index

A

accuracy, 25, 55, 56, 72, 102
achlorophyllous, 137
acid, 11, 116
acid fuchsin, 151
acid rain, 6, 69
Actinomycete, 135
activity, 22
aerobic, 18, 96, 117, 135–139, 176
agricultural crops, 10
alkaline, 11, 87, 95, 105, 117, 122, 125,
    151, 152
aluminum, 9, 10, 47, 52, 69, 71, 76, 83,
    84, 91, 92, 101, 106, 109, 116,
    117, 119, 123, 126, 151, 152,
    155, 158
Amazon, 13
ammonium, 22, 70, 87, 88, 95, 96, 105–
    108, 110, 122, 125
Anabaena, 138
anaerobic, 18, 96, 116, 135, 136, 139
apatite, 101, 103, 104
arbuscular mycorrhizas (AM), 142
arbutoid, 140, 142
Archaea, 135
arid, 84, 85, 87
asexual spores, 137, 138
autoanalyzer, 108, 110, 125
Azolla, 138
Azotobacter, 136

B

bacteria, 133–135, 138, 139, 174, 176
base, 87
base cations, 86, 117
biodiversity, 133, 154
biological, 22

biological activity, 18, 21, 134, 154
biomass, 110, 134, 139, 140
blue-green algae, 138
Bouyoucos hydrometer, 37, 55
buffer capacity, 107
bulk density, 20, 50

C

calcium, 11, 14, 26, 33, 49, 53, 71, 78,
    84, 88, 101, 106, 108, 109,
    124
capillary, 60, 74
carbon, 4, 7, 9, 12, 14, 43, 67, 105, 108,
    111, 121–124, 133, 138, 158
Carolina, 106, 109
cation exchange capacity (CEC), 11, 72,
    82
cations, 87
cellulose, 111, 134, 136, 137
charcoal, 13, 105
chlorazol black E, 151
clay, 10–13, 22, 23, 46, 51, 56, 57, 61,
    69, 82, 85, 86, 122
clear-cut, 116
climate change, 7, 8, 14
climate risk, 6
Clod method, 50
CO_2 evolution, 154, 158
coarse fraction volume (CFV), 49
coefficient of variation, 25, 26
cohesion, 60
collapse, 3, 7, 14
colorimetric method, 102
combustion, 7, 112, 124
complexed, 34, 117
copper, 13, 122, 123, 125
core method, 50
Cyanobacteria, 136, 138

D

decay, 168
decomposition, 69, 134, 154
  rates, 69, 134, 154
Decomposition Rates, 159, 167
deforestation, 3, 4, 12, 14
degradation, 5, 6, 11, 111
degrees of freedom, 26
denitrification, 136
Desilication, 10
diazotrophs, 138
dilution plates, 139
disturbance, 5, 8–10, 20, 21, 69
diversity, 7, 134, 149, 169, 170
DNA, 100, 134, 136, 137, 146, 150, 170–
  175
Dokuchaev, 9
drought, 3, 5, 6, 9, 69
dry ashing, 121, 129
Dry oxidation, 124
drying, 21, 23, 38, 62, 96, 113, 129, 157,
  159
Dumas technique, 124

E

Easter Island, 4
ecological footprint, 6
ecosystem services, 6–9, 14
ectendomycorrhiza, 140, 142
Ectomycorrhizas (EM), 140
electrical conductivity, 46, 78, 79, 89
electrometric method, 72
eluviations, 10
endospores, 135, 136
energy state, 59
enzyme, 68, 111, 134, 154, 173, 174
ericoid, 140, 142
erosion, 3, 9, 12, 43
error, 18, 19, 27, 210
  Calculation, 19
  measurement, 19
  sampling, 18, 19, 27, 210
  selection, 19
essential elements, 67, 100
Excavation method, 50
Exchangeable acidity, 91
exchangeable acidity, 87
externalities, 12

F

facultative, 136
FastDNA Spin Kit, 172
feldspar, 52, 85
fertilizer, 31, 69, 96
fixation, 11, 100, 136
floor, 31, 111
Fluorescent stains, 139
foliar analyses, 29
Forest floor, 49
forest management practices, 111
fragmenters, 176
Freezing, 22
fruiting, 30, 142, 147, 150
fumigation, 140
fungi, 133, 134, 138–140, 142, 143, 174,
  176
fusion, 103

G

gas chromatograph (GC), 155
gel electrophoresis, 173–175
geophagy, 13
Glinka, 9
gram, 136
  negative, 136
  positive, 136
gravel, 23, 51
gravimetric, 46, 63
gravitational, 59, 61
Gravity potential, 60
growth, 6, 9, 11, 12, 14, 18, 29, 31, 67,
  68, 70, 71, 76, 100, 106, 117,
  133
guano, 101
Gypsum blocks, 46

H

heavy metal, 116
high specific area, 52
high surface charge, 52
Hilgard, 9
human health, 12, 67, 69
hydrated radius, 85, 86
hydrogen bonding, 58
hydrogen ion activity, 70
Hygroscopic, 59

hyphae, 137, 138, 140, 142, 149, 152
hysteresis, 63

I

Ideal Gas Law, 156
important, 142
inductively coupled plasma spectroscopy
        (ICP), 103
infrared gas analyzer (IRGA), 155
invertebrates, 133, 134
iodine, 67
ion chromatography, 103
iron, 10, 11, 13, 52, 53, 83, 101, 106,
        109, 112, 114, 116, 117, 119,
        123, 126, 136
isotopic dilution, 107

J

Jenny, 9

K

kinetic energy, 59
Kjeldahl, 31, 34, 121, 123, 125

L

land uses, 4
landscape, 3, 86
Lawes, 101
leaching, 4, 7, 11, 14, 30, 69, 71, 117
leaf tissues, 31
legacy, 6, 12
legume, 136, 142
Liebig, 101
lignin, 136, 137
litter, 111, 117, 134, 154, 159, 176
litterbag, 159
lyotropic series, 87
lysis, 146

M

magnesium, 13, 53, 84, 90, 124
manganese, 13, 76, 112, 118, 119, 125
mantle, 140
Marbut, 9
matric potential, 60, 62
Maya, 5

mean, 5, 8, 12, 24, 25, 30, 52, 62, 69,
        73, 77, 78, 97, 135, 175
Mechanical Analysis, 50
Mehlich, 106, 109
metabolic, 68
metagenome, 176
metatranscriptomes, 176
methane, 136
microarthropods, 183
microbe, 107, 116, 117, 133
microfauna, 176
microorganisms, 176
mineral, 8, 33, 46, 48, 67, 83, 84, 86, 95,
        97, 102, 104, 110, 111, 117,
        146
mineralization, 96, 105
mining, 101, 116
moisture, 75
moisture release curves, 63
molybdenum-blue methods, 103
morphotype, 142, 149
mycorrhizas, 134, 137, 142, 149, 151

N

Natural Resources Conservation Service
        (NRCS), 8
Nazca, 4, 12
nematode, 183
Nernst equation, 72
Neutron, 47
nitrate, 22, 34, 75, 95, 107, 124, 125
nitrification, 136
nitrogen, 6, 14, 31, 33, 67, 75, 96, 100,
        111, 112, 121, 136, 138
nodules, 101, 136
Nostoc, 138
nucleic, 68
nucleic acid, 100, 171, 174
NucleoSpin Soil DNA Isolation Kit, 172
nucleotides, 172, 174
number of samples, 18, 26, 27, 129
nutrient status, 29

O

obligate, 137
occlusion, 117
orchid, 140, 142

organic matter, 7, 10–13, 23, 43, 47, 48,
        50, 53, 69, 72, 85, 91, 104,
        105, 111, 112, 133, 135, 147
organism, 3, 13, 117, 137, 139
osmotic potential, 79
oxidation, 34, 112, 121, 123
        dry, 34, 112, 121
        wet, 112, 121, 123
oxygen, 10, 67, 111, 112, 124, 125, 135,
        138

P

parent material, 86
Particle Size Analysis, 50
pathogens, 135, 138
peds, 43
perchloric acids, 121
pH, 10, 70–72, 76, 82, 83, 87, 91, 101,
        105, 106, 116, 135, 146, 171
phenoldisulfonic acid, 34
phosphates, 29, 102, 104, 106–109
phosphorus, 11, 13, 101, 103, 107, 117,
        124
photosynthesis, 68, 116, 138
pipet method, 55
plant sampling, 16
Podzolization, 10
poll, 116
pollution, 12, 67, 69
polymerase, 172, 174
polymerase chain reaction (PCR), 139
population, 5, 17, 19, 133, 135, 140, 174
potassium, 13, 33, 96, 100, 110, 112,
        115, 122, 125, 130
potential energy, 59, 61
PowerSoil, PowerMax and UltraClean
        soil DNA isolation, 172
precision, 18, 25, 26
pressure plate, 62, 63
pressure potential, 62
primers, 170, 172–174
protozoans, 134, 177
pyrosequencing, 176

Q

QIAGEN Multiplex PCR Kit, 173
Quality Assurance, 88
Quality Control (QA/QC), 88
Quantitative PCR (qPCR), 173

R

radiation, 47, 50, 84
radioactive tracer, 84, 107
reduction, 12, 68, 103, 136
replication, 172, 174
resilience, 6, 10–12
resins, 103, 106
        anion, 106
        ion-exchange, 103
restriction fragment length polymorphism
        (RFLP), 139
ribosomal DNA (rDNA), 174
RNA, 100, 136, 146, 173–175
rocks, 3, 12, 23, 24, 50, 95, 103, 138
roots, 23, 140, 142, 152, 159, 161–163
        coarse, 23, 161–163
        fine, 140, 142, 152, 159
rotifers, 177

S

salinity, 9, 46, 75–77, 79
salts, 3, 11, 12, 37, 72, 75, 77, 78, 87,
        88, 122, 124
samples, 26
sampling design, 16, 17, 31, 150
sand, 51, 53, 57
saprophytes, 137
saturated paste, 77, 79, 80
sedimentary, 3, 101
sedimentation, 52, 54–56
selenium, 122, 123
senesced tissue, 33
sequencing, 139, 147, 170, 172, 174–176
sexual spores, 137, 138
sheath, 140
sieving, 52, 53
silt, 51, 57
simple random sample, 17
slime molds, 138
soda lime technique, 155
sodic, 11
sodium, 10, 13, 36, 53, 56, 84, 86, 105,
        109, 117, 118, 122, 125, 128,
        131, 148, 171
soil animal, 18
soil bulk, 48
soil bulk density, 43, 49
soil health, 11, 12, 14, 133

soil moisture, 43, 46, 114
soil pores, 46, 62, 71, 84
soil respiration, 134, 154, 155, 158
soil water potential, 37, 43, 62
soil water retention curves, 63
SoilMaster DNA Extraction Kit, 171
sporocarps, 150
stain, 135, 136, 151, 175
standard deviation, 25
standard error, 25, 26
statistical, 26
stem tissues, 32
sterilization, 139
storage, 7, 14, 22–24, 70, 96, 98, 133, 134, 142, 173
strengite, 102, 104
substrate utilization profile, 134
sulphur, 136
Sumerian, 11, 14
SurePrep^{TM} Soil DNA Isolation Kit, 172
surface tension, 58, 61, 62
sustainable, 5
systematic sample, 18

T

tardigrades, 177

termites, 176
terra preta, 7
time domain reflectometry (TDR), 27
tipping points, 8
titration, 112, 115, 123, 124
transcriptome, 176
transport, 4, 155
Trypan blue, 151, 152

V

valence, 85, 86, 88, 90
variance, 26
variscite, 102, 104
volumetric, 45, 128, 129
vulnerability index, 6

W

water potential, 61, 62
wood lice, 176
woody debris, 33

Z

zinc, 13, 125
zoospores, 138

www.ingramcontent.com/pod-product-compliance
Lightning Source LLC
Chambersburg PA
CBHW081358270326
41930CB00015B/3340